Study Guide and Selected Solutions Manual

CHARLES H. CORWIN

EIGHTH EDITION

Introductory Chemistry

CONCEPTS AND CRITICAL THINKING

CHARLES H. CORWIN

Printed in the United States of America
1 17

PEARSON
www.pearsonhighered.com

Pearson Prentice Hall
Upper Saddle River, NJ 07458

Contents

Selected Solutions Manual

for Odd-Numbered Textbook Exercises

Flashcards

Preface

To the Student

This *Study Guide and Selected Solutions Manual* will help you learn chemistry while using the textbook ***Introductory Chemistry Concepts and Connections,* Eighth Edition**. This valuable resource allows you to test your understanding of each topic and discover areas that require additional study. Thus, you will find self-test questions for each objective in the textbook.

There is also a crossword puzzle following the self-test questions to assist you in learning the key terms in each chapter. The answers to the self-test questions, as well as the crossword puzzle, are included at the end of each set of chapter questions of this *Study Guide* and *Selected Solutions Manual.*

The following guidelines will help you achieve maximum success in your introductory chemistry course.

- **First**, pay close attention to the assignments given by your instructor.
- **Second**, read the assigned sections in the text and study the example exercises. Do the end-of-chapter exercises and check your answers in *Appendix I*. If you cannot solve a problem or have difficulty with an exercise, refer to the complete solution at the end of this study guide.
- **Third**, answer the self-test questions in this study guide to diagnose any lack of understanding in the assignments by your instructor. If you answer a question incorrectly, refer to the corresponding section in the textbook.
- **Fourth**, take the practice Self-Test at the end of each assigned chapter in the textbook and check your answers in *Appendix J*. If you answer a question incorrectly, refer to the corresponding section in the textbook.

If you observe these suggestions, this *Study Guide* and *Selected Solutions Manual* will make learning chemistry much more efficient. For those topics that still seem unclear, visit your instructor during office hours. Instructors are happy to help students, especially those who are making a serious effort to do the assignments and learn the material.

If you want to share your chemistry experience, please send an e-mail or drop a note at the address below. I would like to hear about your success as well as any suggestions for improving these study aids.

Charles H. Corwin
Department of Chemistry
American River College
Sacramento, CA 95841
corwinc@arc.losrios.edu

Study Guide

with

Self-Test Questions

and

Crossword Puzzles of Key Terms

EIGHTH EDITION

CHAPTER

1

Introduction to Chemistry

Section 1.1 *Evolution of Chemistry*

1. What four elements composed everything in nature according to the beliefs of the ancient Greeks?
   (a) air, earth, salt, and water
   (b) air, ashes, fire, and water
   (c) air, earth, fire, and water
   (d) smoke, earth, fire, and water
   (e) none of the above

2. Which of the following is a basic step in the scientific method?
   (a) perform an experiment and collect data
   (b) analyze experimental data and propose a hypothesis
   (c) test a hypothesis and state a theory or law
   (d) all of the above
   (e) none of the above

Section 1.2 *Modern Chemistry*

3. In which of the following industries does chemistry play an important role?
   (a) agriculture
   (b) electronics
   (c) paper
   (d) transportation
   (e) all of the above

4. Which of the following is *not* derived from a petrochemical?
   (a) a gold coin
   (b) a plastic toy
   (c) an insecticide
   (d) a synthetic fabric
   (e) a blue dye

Section 1.3 *Learning Chemistry*

5. In a survey by the *American Chemical Society* of entering college students, which science course was considered the most relevant to their daily lives?
   (a) biology
   (b) chemistry
   (c) geology
   (d) physics
   (e) physiology

6. Which of the following is a positive association with chemistry?
   (a) chemistry has biomedical applications
   (b) chemistry involves interesting experiments
   (c) chemistry topics are relevant to our daily lives
   (d) chemistry has many career opportunities
   (e) all of the above

7. Connect the following nine dots using:

   (a) 5 straight, continuous lines.

   (b) 4 straight, continuous lines.

   (c) 3 straight, continuous lines.

Chapter 1 Key Terms

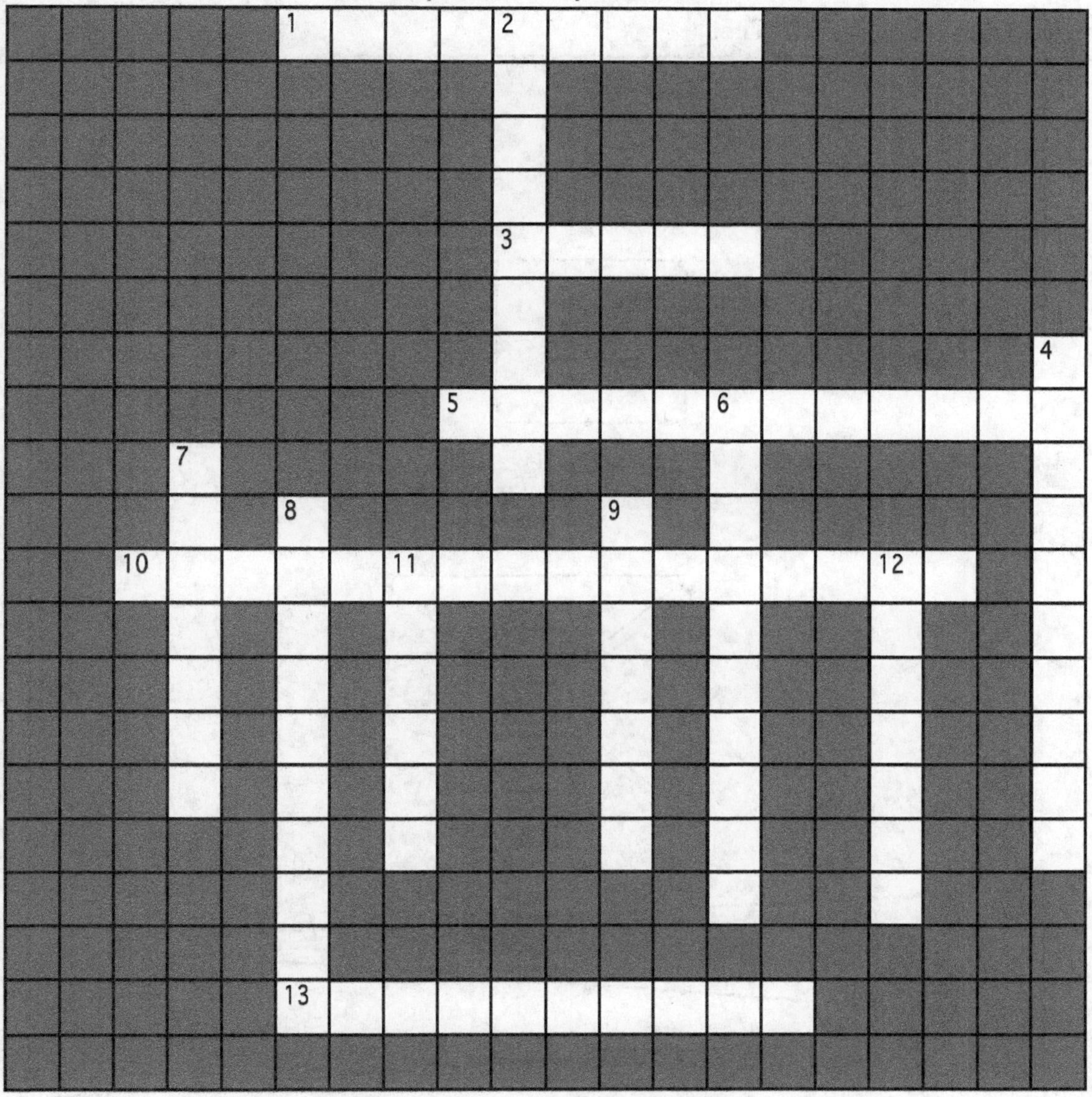

Across

1. the study of the composition of matter
3. chemistry of products that reduces waste
5. the study of biological substances and reactions
10. a systematic investigation of nature (2 words)
13. measurable scientific proposal (2 words)

Down

2. substances not containing carbon
4. a proposal to explain experimental results
6. a scientific procedure for collecting data
7. a pseudoscience for changing lead into gold
8. "chemistry is hard, chemistry is easy"
9. the methodical exploration of nature
11. an extensively tested scientific proposal
12. the chemistry of substances containing carbon

Chapter 1 Answers to Self-Test

1. **c** 2. **d** 3. **e** 4. **a** 5. **b** 6. **e**

7. (a)

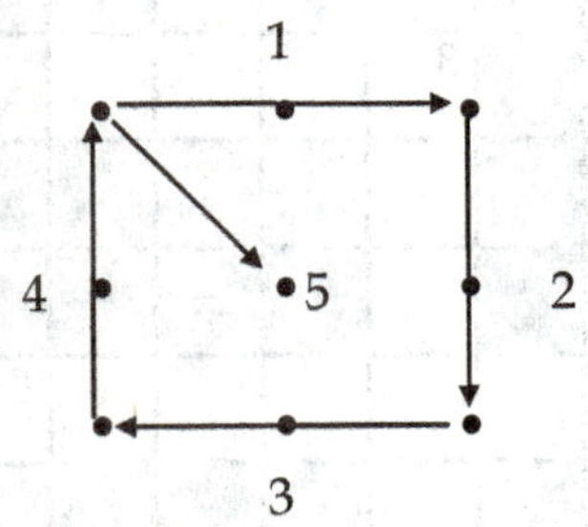

(b)

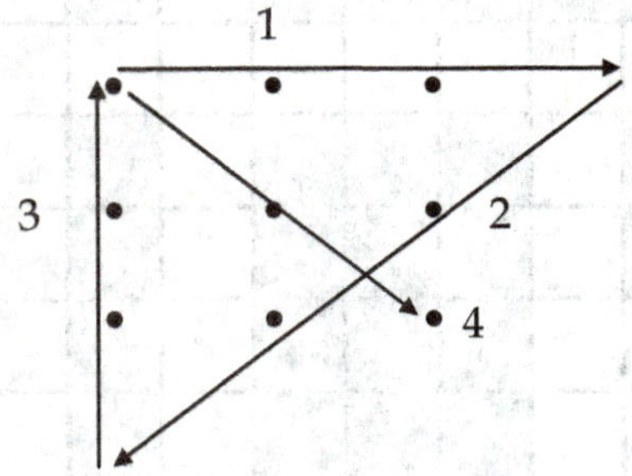

(c)

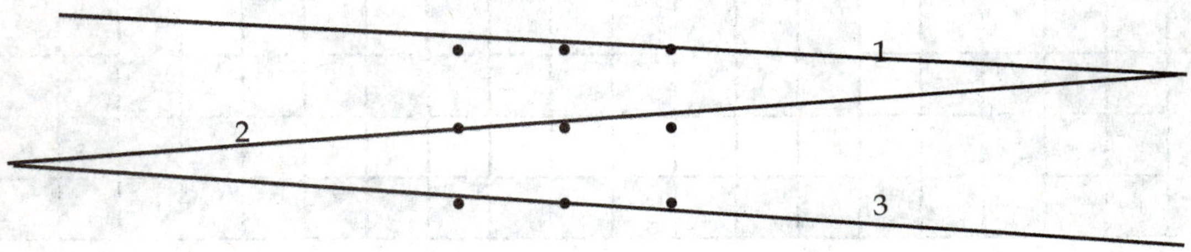

Chapter 1 Answers to Crossword Puzzle

Across

1. chemistry
3. green
5. biochemistry
10. scientific method
13. natural law

Down

2. inorganic
4. hypothesis
6. experiment
7. alchemy
8. perception
9. science
11. theory
12. organic

CHAPTER

PSS

Prerequisite Science Skills

Section PSS.1 *Measurements*

1. Which of the following quantities can be measured exactly?
   (a) length
   (b) mass
   (c) volume
   (d) time
   (e) no quantity can be measured exactly

2. Which of the following metric rulers has the *least* uncertainty?
   (a) Ruler A, ± 1 cm
   (b) Ruler B, ± 0.1 cm
   (c) Ruler C, ± 0.5 cm
   (d) Ruler D, ± 0.05 cm
   (e) all rulers have the same uncertainty

3. Which of the following balances has the *least* uncertainty?
   (a) analytical balance, ± 0.0001 g
   (b) electronic balance, ± 0.001 g
   (c) beam balance, ± 0.01 g
   (d) platform balance, ± 0.1 g
   (e) all balances have the same uncertainty

4. Which of the following instruments has the *least* uncertainty?
   (a) graduated cylinder, ± 0.5 mL
   (b) buret, ± 0.05 mL
   (c) syringe, ± 0.001 mL
   (d) pipet, ± 0.1 mL
   (e) all instruments have the same uncertainty

Section PSS.2 *Significant Digits*

5. How many significant digits are in the length measurement 20.05 cm?
   (a) 1
   (b) 2
   (c) 3
   (d) 4
   (e) none of the above

6. How many significant digits are in the mass measurement 0.0550 g?
   (a) 2
   (b) 3
   (c) 4
   (d) 5
   (e) none of the above

7. How many significant digits are in the volume measurement 5050.0 mL?
   (a) 2
   (b) 3
   (c) 4
   (d) 5
   (e) none of the above

Section PSS.3 *Rounding Off Nonsignificant Digits*

8. Round off the following measurement to three significant digits: 107.500 g.
   (a) 107 g
   (b) 108 g
   (c) 107.0 g
   (d) 107.5 g
   (e) none of the above

9. Round off the following measurement to three significant digits: 0.03580 cm.
   (a) 0.03 cm
   (b) 0.035 cm
   (c) 0.0358 cm
   (d) 0.03580 cm
   (e) none of the above

10. Round off the following measurement to three significant digits: 3265 mL.
   (a) 326 mL
   (b) 327 mL
   (c) 3260 mL
   (d) 3270 mL
   (e) none of the above

Section PSS.4 *Adding and Subtracting Measurements*

11. Add 7.77 g to 51.665 g and round off the answer.
    (a) 59.0 g
    (b) 59.4 g
    (c) 59.43 g
    (d) 59.44 g
    (e) 59.436 g

12. Subtract 2.25 cm from 11.5 cm and round off the answer.
    (a) 9.0 cm
    (b) 9.2 cm
    (c) 9.20 cm
    (d) 9.25 cm
    (e) 9.3 cm

Section PSS.5 *Multiplying and Dividing Measurements*

13. Multiply 2.505 m times 1.75 m and round off the answer.
    (a) 4.0 m^2
    (b) 4.00 m^2
    (c) 4.38 m^2
    (d) 4.384 m^2
    (e) 4.40 m^2

14. Divide 124.7 miles (mi) by 2.25 hours (h) and round off the answer.
    (a) 50 mi/h
    (b) 55 mi/h
    (c) 55.0 mi/h
    (d) 55.4 mi/h
    (e) 55.5 mi/h

Section PSS.6 *Exponential Numbers*

15. Express 100,000,000,000,000,000,000 as an exponential number.
    (a) 1×10^{-20}
    (b) 1×10^{-18}
    (c) 1×10^{18}
    (d) 1×10^{20}
    (e) none of the above

16. Express 0.000 000 000 000 000 000 000 000 000 000 010 as an exponential number.
    (a) 1.0×10^{-33}
    (b) 1.0×10^{-32}
    (c) 1.0×10^{-33}
    (d) 1.0×10^{32}
    (e) none of the above

17. Express the exponential number 5×10^{-9} as an ordinary number.
    (a) 5,000,000,000
    (b) 500,000,000
    (c) 0.000 000 005
    (d) 0.000 000 000 5
    (e) none of the above

Section PSS.7 *Scientific Notation*

18. One gram of carbon contains 50,200,000,000,000,000,000,000 atoms. Express the number of carbon atoms in scientific notation.
    (a) 5.02×10^{21}
    (b) 5.02×10^{22}
    (c) 5.02×10^{23}
    (d) 50.2×10^{21}
    (e) none of the above

19. A carbon atom has a mass of 0.000 000 000 000 000 000 000 019 9 g. Express the mass in scientific notation.
    (a) 1.99×10^{-21} g
    (b) 1.99×10^{-22} g
    (c) 1.99×10^{-23} g
    (d) 1.99×10^{-24} g
    (e) none of the above

20. A carbon atom has a radius of 7.7×10^{-9} cm. Express the atomic radius as an ordinary number.
    (a) 0.000 000 000 77 cm
    (b) 0.000 000 007 7 cm
    (c) 0.000 000 077 cm
    (d) 0.000 000 77 cm
    (e) none of the above

PSS Key Terms

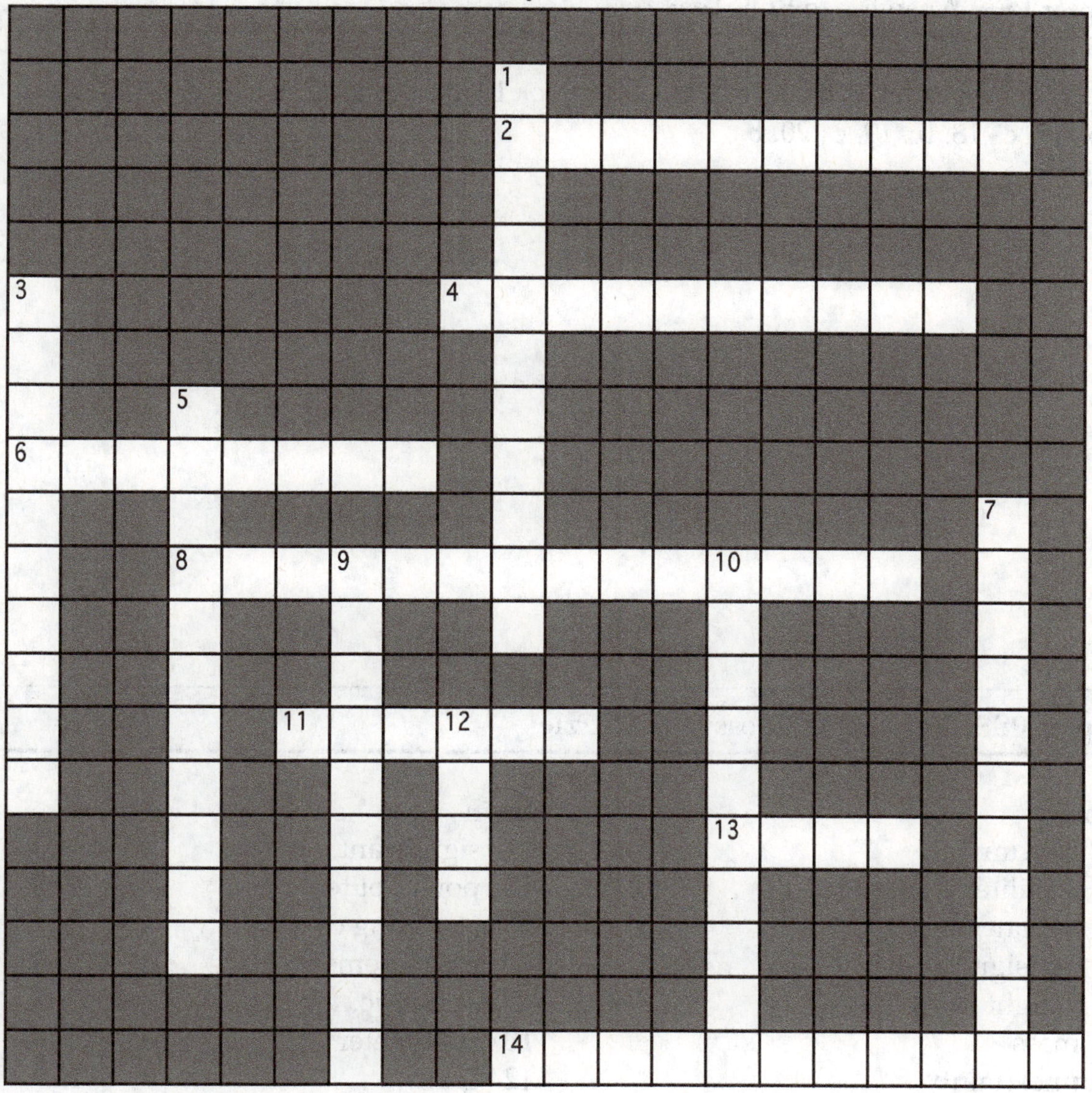

Across

2. a device for obtaining a measurement
4. a metric unit of volume
6. a superscript number
8. the digits often displayed in a calculator
11. the force exerted by gravity on an object
13. the quantity of matter in an object
14. the inexactness of a measurement

Down

1. the digits obtained from a measurement
3. 10^n (3 words)
5. eliminating nonsignificant digits (2 words)
7. a numerical value followed by units
9. a type of notation such as 3.35×10^7
10. a metric unit of length
12. a metric unit of mass

Chapter PSS Answers to Self-Test

1. **e** 2. **d** 3. **a** 4. **c** 5. **d** 6. **b** 7. **d** 8. **b** 9. **c** 10. **d** 11. **d** 12. **e** 13. **c** 14. **d** 15. **d** 16. **b** 17. **c** 18. **b** 19. **c** 20. **b**

Chapter PSS Answers to Crossword Puzzle

Across

2. instrument
4. milliliter
6. exponent
8. nonsignificant
11. weight
13. mass
14. uncertainty

Down

1. significant
3. power of ten
5. rounding off
7. measurement
9. scientific
10. centimeter
12. gram

CHAPTER

2

The Metric System

Section 2.1 *Basic Units and Symbols*

1. Which of the following are the basic units and symbols for the metric system?
   (a) centimeter (cm), gram (g), liter (L)
   (b) centimeter (cm), gram (g), milliliter (mL)
   (c) meter (m), gram (g), liter (L)
   (d) meter (m), gram (g), liter (l)
   (e) meter (m), gram (gm), liter (L)

2. What is the symbol for the metric unit microliter?
   (a) cL
   (b) mL
   (c) ML
   (d) μL
   (e) none of the above

Section 2.2 *Metric Conversion Factors*

3. According to the metric system, 1 m = 10 ______.
   (a) cm
   (b) dm
   (c) mm
   (d) nm
   (e) μm

4. According to the metric system, 1 g = 100 ______.
   (a) kg
   (b) dg
   (c) cg
   (d) mg
   (e) μg

5. Which of the following is a unit factor corresponding to the unit equation: 1 m = 39.4 in.?
   (a) 1 m/1 in.
   (b) 1 m/39.4 in.
   (c) 39.4 in./39.4 m
   (d) 1 in./39.4 m
   (e) none of the above

6. Which of the following is a unit factor corresponding to the unit equation: 1 m = 100 cm?
   (a) 1 m/1 cm
   (b) 100 m/100 cm
   (c) 1 cm/100 m
   (d) 100 cm/1 m
   (e) none of the above

Section 2.3 *Metric–Metric Conversions*

7. What is the three-step sequence for the unit analysis method of problem solving?
   (a) 1–relevant known value, 2–unknown units, 3–unit factor
   (b) 1–unknown units, 2–unit factor, 3–relevant known value
   (c) 1–unknown units, 2–relevant known value, 3–unit factor
   (d) 1–unit factor, 2–unknown units, 3–relevant known value
   (e) 1–unit factor, 2–relevant known value, 3–unknown units

8. What is the length in meters of a 20.0-mL test tube that measures 15.0 cm?
   (a) 0.150 m
   (b) 0.200 m
   (c) 1.50 m
   (d) 1500 m
   (e) 2000 m

9. A tin atom has a radius of 1.41×10^{-8} cm. What is the radius in nanometers?
   (a) 1.41×10^{-16} nm
   (b) 1.41×10^{-10} nm
   (c) 1.41×10^{-4} nm
   (d) 1.41×10^{-1} nm
   (e) 1.41×10^{5} nm

Section 2.4 *Metric–English Conversions*

10. Which of the following English–metric equivalents is correct?
   (a) 1 in. = 2.54 cm
   (b) 1 lb = 454 g
   (c) 1 qt = 946 mL
   (d) all of the above
   (e) none of the above

11. A stainless steel cylinder weighs 0.892 lb. What is the mass in kilograms?
   (a) 0.405 kg
   (b) 4.05 kg
   (c) 40.5 kg
   (d) 4050 kg
   (e) 405,000 kg
12. A glass bottle has a volume of 0.750 L. What is the volume in quarts?
   (a) 0.605 qt
   (b) 0.750 qt
   (c) 0.793 qt
   (d) 1.26 qt
   (e) 1.65 qt

Section 2.5 *The Percent Concept*

13. A 1¢ coin minted before 1982 is an alloy of copper and zinc. If the penny contains 2.869 g copper metal and 0.151 g zinc metal, what is the percent of zinc in the coin?
   (a) 5.00%
   (b) 5.27%
   (c) 15.1%
   (d) 20.0%
   (e) 95.0%
14. A 10¢ coin minted after 1971 is an alloy of copper and nickel. If the 10¢ coin has a mass of 2.550 g and is 75.0% copper, what is the mass of nickel in the coin?
   (a) 0.102 g
   (b) 0.340 g
   (c) 0.638 g
   (d) 0.850 g
   (e) 1.91 g

Section 2.6 *Volume by Calculation*

15. A gold foil has a volume of 0.645 cm^3. If the gold foil measures 10.0 cm by 12.5 cm, what is the thickness of the foil?
   (a) 0.000516 cm
   (b) 0.00516 cm
   (c) 0.0516 cm
   (d) 0.516 cm
   (e) 0.806 cm
16. If an automobile has an 115-in.^{3} engine, what is the volume of the engine in cubic centimeters?
   (a) 1.88 cm^3
   (b) 7.01 cm^3
   (c) 292 cm^3
   (d) 742 cm^3
   (e) 1880 cm^3

Section 2.7 *Volume by Displacement*

17. What is the term for the method of determining volume by measuring the increase in water level after an object is completely immersed?
    (a) volume by density
    (b) volume by difference
    (c) volume by displacement
    (d) volume by water level
    (e) none of the above

18. A 200.0-g sample of cobalt metal is dropped into a 100-mL graduated cylinder containing 45.5 mL of water. If the resulting water level is 68.0 mL, what is the volume of the cobalt?
    (a) 12.5 mL
    (b) 22.5 mL
    (c) 32.0 mL
    (d) 54.5 mL
    (e) 113.5 mL

Section 2.8 *The Density Concept*

19. If the density of cobalt metal is 4.51 g/mL, which of the following is an associated unit factor?
    (a) 1 g/1 mL
    (b) 4.51 g/4.51 mL
    (c) 4.51 g/1 mL
    (d) 4.51 mL/1 g
    (e) none of the above

20. If the density of ethyl alcohol is 0.789 g/mL, what is the volume of 35.5 g of ethyl alcohol?
    (a) 2.80 mL
    (b) 4.50 mL
    (c) 28.0 mL
    (d) 45.0 mL
    (e) 280 mL

21. A glass cylinder contains four separate liquid layers: mercury (d = 13.6 g/mL), chloroform (d = 1.49 g/mL), acetic acid (d = 1.05 g/mL), ether (d = 0.708 g/mL). If a marble (d = 3.05 g/mL) is added to the cylinder, where does it come to rest?
    (a) on top of the mercury layer
    (b) on top of the chloroform layer
    (c) on top of the acetic acid layer
    (d) on top of the ether layer
    (e) on the bottom of the cylinder

Section 2.9 *Temperature*

22. What is the freezing point and boiling point of water on the Celsius scale?
    (a) 0 °C and 100 °C
    (b) 0 °C and 212 °C
    (c) 32 °C and 100 °C
    (d) 32 °C and 212 °C
    (e) 273 °C and 373 °C

23. Given that methanol freezes at –94 °C, what is the freezing point temperature on the Kelvin scale?
    (a) -367 K
    (b) -179 K
    (c) -94 K
    (d) 179 K
    (e) 367 K

Section 2.10 *The Heat Concept*

24. Which of the following may express the total amount of heat energy in a sealed, insulated chamber?
    (a) 25 °C
    (b) 77 °F
    (c) 298 K
    (d) 125 kcal
    (e) all of the above

25. How many kilocalories of heat are required to raise 250.0 g of liquid water from 20.0 °C to 75.0 °C?
    (a) 0.220 kcal
    (b) 4.55 kcal
    (c) 5.00 kcal
    (d) 13.8 kcal
    (e) 18.8 kcal

26. A 50.0-g sample of aluminum releases 420.0 calories when cooled from 100.0 °C to 60.0 °C. What is the specific heat of the metal?
    (a) 0.168 cal/(g × °C)
    (b) 0.210 cal/(g × °C)
    (c) 0.840 cal/(g × °C)
    (d) 4.76 cal/(g × °C)
    (e) 21.0 cal/(g × °C)

Unit Analysis Problem Solving – Show all work for the following.

27. If the radius of a nickel atom is 0.125 nm, what is the atomic radius of a nickel atom in picometers (pm)? (Given: $1 \text{ m} = 1 \times 10^{12} \text{ pm}$)

28. If a roll of nickel coins has a mass of 205 g, what is the mass of the roll of nickels in ounces? (Given: 1 lb = 16 ounces)

Chapter 2 Key Terms

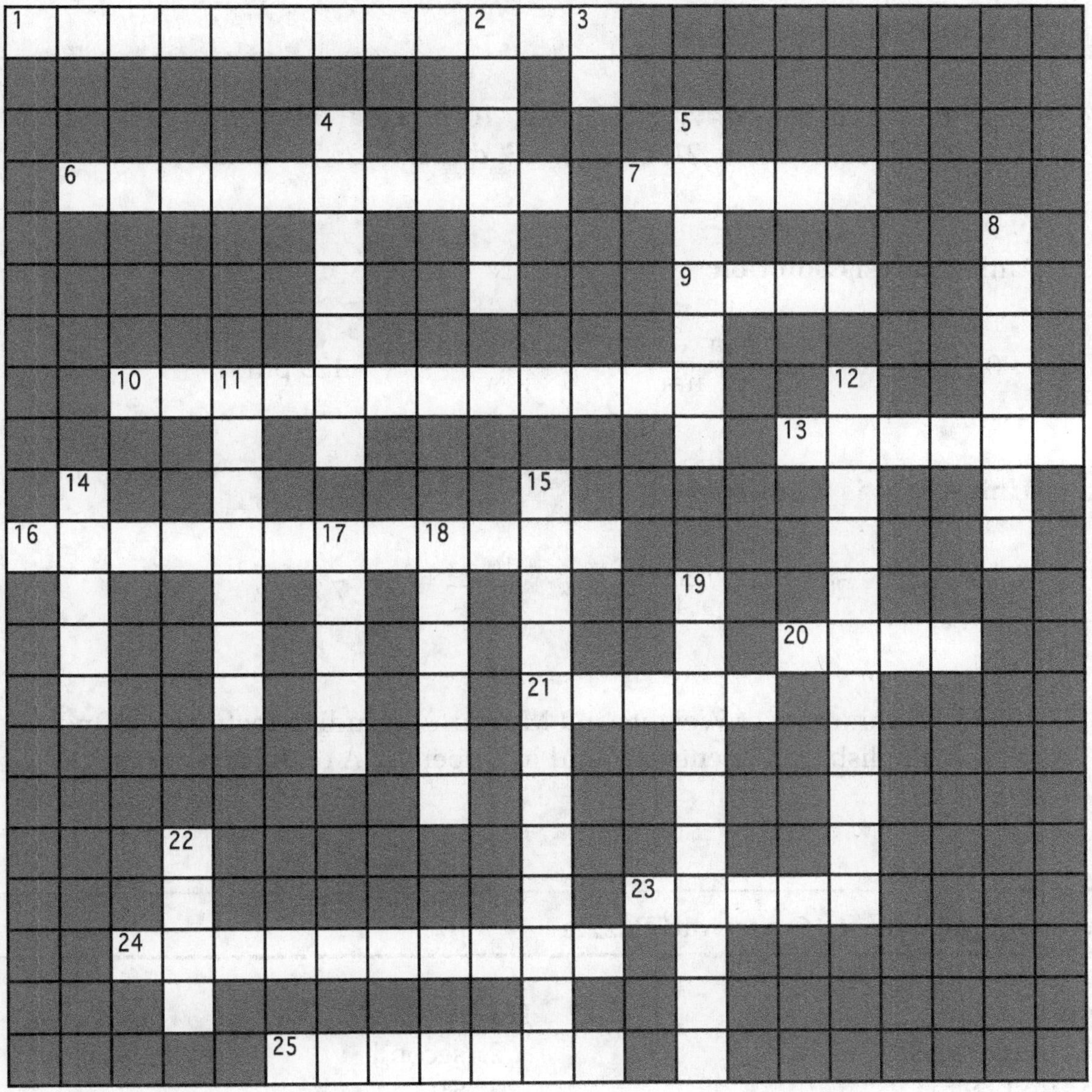

Across

1. a problem-solving method (2 words)
6. the ratio of two equivalent quantities (2 words)
7. a length approximately equal to a yard
9. abbreviation for specific gravity
10. value for water is 1.00 cal/(g x °C) (2 words)
13. ice melts at 273 on this temperature scale
16. a method for determining volume
20. basic unit of mass in metric system
21. 1 L = 1000 mL is an _______ equivalent
23. basic unit of heat in the International System
24. ice melts at 32° on this temperature scale
25. parts per 100 parts

Down

2. basic unit of time in metric system
3. abbreviation for the International System
4. the heat required to raise 1 g of water 1 °C
5. mass per unit volume
8. ice melts at 0° on this temperature scale
11. a nondecimal system of measurement
12. average energy of molecules in motion
14. basic unit of volume in metric system
15. states two equivalent quantities (2 words)
17. 1 mL = 1 _________ centimeter
18. a decimal system of measurement
19. the inverse of a fraction
22. the flow of energy from a hot to cold object

Chapter 2 Answers to Self-Test

1. **c** 2. **d** 3. **b** 4. **c** 5. **b** 6. **d** 7. **c** 8. **a** 9. **d** 10. **d** 11. **a** 12. **c** 13. **a** 14. **c** 15. **b** 16. **e** 17. **c** 18. **b** 19. **c** 20. **d** 21. **a** 22. **a** 23. **d** 24. **d** 25. **d** 26. **b**

27. **Unit Analysis Solution:**

$$0.125\ \cancel{\text{nm}} \times \frac{1\ \cancel{\text{m}}}{1 \times 10^{9}\ \cancel{\text{nm}}} \times \frac{1 \times 10^{12}\ \text{pm}}{1\ \cancel{\text{m}}} = 125\ \text{pm}$$

28. **Unit Analysis Solution:**

$$205\ \cancel{\text{g}} \times \frac{1\ \cancel{\text{lb}}}{454\ \cancel{\text{g}}} \times \frac{16\ \text{oz}}{1\ \cancel{\text{lb}}} = 7.22\ \text{oz}$$

(Note: A summary of Weights and Measures including metric units and English equivalents is found in Appendix A of the textbook.)

Chapter 2 Answers to Crossword Puzzle

Across
1. unit analysis
6. unit factor
7. meter
9. spgr
10. specific heat
13. Kelvin
16. displacement
20. gram
21. exact
23. joule
24. Fahrenheit
25. percent

Down
2. second
3. SI
4. calorie
5. density
8. Celsius
11. English
12. temperature
14. liter
15. unit equation
17. cubic
18. metric
19. reciprocal
22. heat

CHAPTER

3

Matter and Energy

Section 3.1 *Physical States of Matter*

1. Which physical state demonstrates the greatest motion of individual particles?
   (a) solid state
   (b) liquid state
   (c) gaseous state
   (d) the motion of particles cannot be predicted for a physical state
   (e) the motion of particles is the same for each physical state

2. Which of the following describes a substance in the solid physical state?
   (a) The substance has a fixed shape.
   (b) The substance has a fixed volume.
   (c) Particles have an ordered arrangement.
   (d) Particles vibrate in fixed positions.
   (e) all of the above

Section 3.2 *Elements, Compounds, and Mixtures*

3. Which of the following has indefinite, variable properties and can be separated into two or more pure substances by physical methods?
   (a) compound
   (b) element
   (c) mixture
   (d) all of the above
   (e) none of the above

4. Which of the following has definite properties and can be broken down into two or more pure substances by a chemical reaction?
   (a) compound
   (b) element
   (c) mixture
   (d) all of the above
   (e) none of the above

5. Sterling silver is an alloy of copper and silver with the percentage of copper always equal to 7.50%. Classify sterling silver as one of the following.
   (a) compound
   (b) element
   (c) heterogeneous mixture
   (d) homogeneous mixture
   (e) none of the above

Section 3.3 *Names and Symbols of the Elements*

6. Which of the following elements is one of the ten most abundant in Earth's crust, oceans, and atmosphere?
   (a) oxygen
   (b) silicon
   (c) aluminum
   (d) titanium
   (e) all of the above

Section 3.4 *Metals, Nonmetals, and Semimetals*

7. Which of the following is a general characteristic of a nonmetal?
   (a) brittle in the solid state
   (b) nonconductor of heat
   (c) low density
   (d) low melting point
   (e) all of the above

8. Which of the following is a property of a metal?
   (a) dull appearance
   (b) reacts with nonmetals
   (c) reacts with other metals
   (d) electrical insulator
   (e) all of the above

9. Which of the following elements is an example of a semimetal?
   (a) H
   (b) C
   (c) Fe
   (d) Ge
   (e) Sn

10. Use a periodic table and predict which of the following elements is a gas at 25 °C and 1 atmosphere pressure.
   (a) N
   (b) Na
   (c) Nb
   (d) Nd
   (e) none of the above

11. Use a periodic table and predict which of the following elements is a liquid at 25 °C and 1 atmosphere pressure.
   (a) B
   (b) Ba
   (c) Be
   (d) Br
   (e) none of the above

12. Use a periodic table and predict which of the following elements is a solid at 25 °C and 1 atmosphere pressure.
   (a) H
   (b) He
   (c) Hg
   (d) Ho
   (e) none of the above

Section 3.5 *Compounds and Chemical Formulas*

13. In which of the following compounds are the elements present in a definite proportion by mass?
   (a) H_2O
   (b) CO_2
   (c) N_2O_4
   (d) NaCl
   (e) all compounds have constant composition

14. Ammonium phosphate is used in fertilizer to replenish nitrogen to the soil. If the formula is $(NH_4)_3PO_4$, what is the total number of atoms in one molecule?
   (a) 13
   (b) 16
   (c) 18
   (d) 20
   (e) none of the above

Section 3.6 *Physical and Chemical Properties*

15. Which of the following is an example of a physical property?
    (a) color
    (b) crystalline form
    (c) melting point
    (d) physical state
    (e) all of the above

16. Which of the following is an example of a chemical property?
    (a) density
    (b) ductility
    (c) electrical conductivity
    (d) flammability
    (e) hardness

17. Which of the following is an example of a chemical property of methyl alcohol?
    (a) methyl alcohol and sulfuric acid yield methyl ether
    (b) methyl alcohol in animals produces blindness
    (c) methyl alcohol and sodium metal release hydrogen gas
    (d) methyl alcohol and formic acid produce rum flavor
    (e) all of the above

Section 3.7 *Physical and Chemical Changes*

18. Which of the following observations is evidence of a physical change?
    (a) condensation
    (b) crystallization
    (c) sublimation
    (d) vaporization
    (e) all of the above

19. Which of the following observations is evidence of a chemical change?
    (a) producing bubbles after mixing solutions
    (b) producing a precipitate after mixing solutions
    (c) producing heat after mixing solutions
    (d) producing a color change after mixing solutions
    (e) all of the above

Section 3.8 *Conservation of Mass*

20. If 558 kg of powdered iron react with powdered sulfur to produce 879 kg of iron sulfide, what is the mass of reacting sulfur?
    (a) 321 kg
    (b) 558 kg
    (c) 879 kg
    (d) 1437 kg
    (e) impossible to predict from the given information

Section 3.9 *Potential and Kinetic Energy*

21. Which of the following is an example of kinetic energy?
    (a) colliding molecules
    (b) moving roller coaster
    (c) swinging in a swing
    (d) rolling wheels
    (e) all of the above

22. Which of the following physical states possesses the greatest attraction between individual particles?
    (a) solid state
    (b) liquid state
    (c) gaseous state
    (d) plasma state
    (e) the attraction is the same for each physical state

23. According to the kinetic theory, temperature measures which of the following?
    (a) molecular attraction
    (b) molecular mass
    (c) molecular motion
    (d) molecular size
    (e) none of the above

Section 3.10 *Conservation of Energy*

24. A nuclear power plant uses radioactive uranium to convert water to steam; steam then drives a turbine that turns a generator for electricity. What are the two forms of energy represented by uranium and the turbine?
    (a) chemical and mechanical energy
    (b) chemical and heat energy
    (c) electrical and mechanical energy
    (d) nuclear and heat energy
    (e) nuclear and mechanical energy

25. According to the law of conservation of mass and energy, which of the following is true of the detonation of a nuclear bomb?
    (a) the mass of the bomb and the products are identical
    (b) the energy of the bomb and the products are identical
    (c) a small amount of matter is converted into energy
    (d) all of the above
    (e) none of the above

Chemical Symbols of the Elements

26. Provide the chemical symbol for the name of the following elements.

(a) aluminum –
(b) antimony –
(c) argon –
(d) arsenic –
(e) barium –
(f) beryllium –
(g) bismuth –
(h) boron –
(i) bromine –
(j) cadmium –
(k) calcium –
(l) carbon –
(m) chlorine –
(n) chromium –
(o) cobalt –
(p) copper –
(q) fluorine –
(r) germanium –
(s) gold –
(t) helium –
(u) hydrogen –
(v) iodine –
(w) iron –
(x) krypton –

27. Write the name of the element for each of the following chemical symbols.

(a) Pb –
(b) Li –
(c) Mg –
(d) Mn –
(e) Hg –
(f) Ne –
(g) Ni –
(h) N –
(i) O –
(j) P –
(k) Pt –
(l) K –
(m) Ra –
(n) Se –
(o) Si –
(p) Ag –
(q) Na –
(r) Sr –
(s) S –
(t) Te –
(u) Sn –
(v) Ti –
(w) Xe –
(x) Zn –

Chapter 3 Key Terms

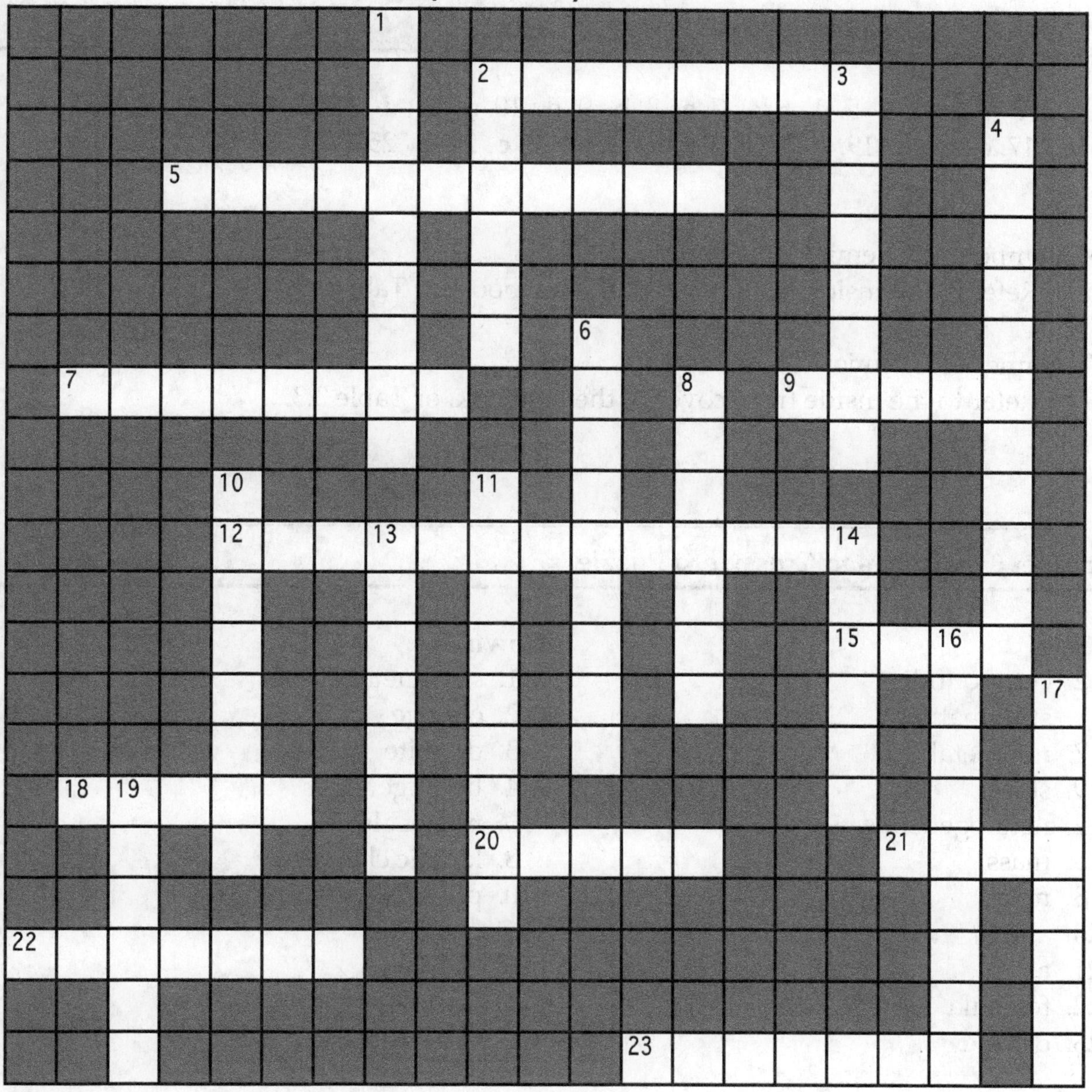

Across

2. it can be broken down into elements
5. a direct change of state from a solid to a gas
7. an element that reacts with metals
9. it can be solid, liquid, or gas
12. a mixture with variable properties
15. it cannot be created nor destroyed
18. a shiny, high-density element
20. a homogeneous mixture of metals
21. a particle that represents an element
22. abbreviation for the name of a compound
23. the property of being drawn into a wire

Down

1. a metalloid
2. a chemical reaction is a chemical ______
3. all compounds have this type of composition
4. a mixture with consistent properties
6. the property of being hammered into a foil
8. the energy of a moving particle
10. a change that does not alter composition
11. the stored energy of chemical composition
13. it cannot be broken down by chemical reaction
14. abbreviation for the name of an element
16. matter having a constant composition
17. a property that involves another substance
19. it can be chemical, mechanical, or electrical

Chapter 3 Answers to Self-Test

1. **c** 2. **e** 3. **c** 4. **a** 5. **d** 6. **e** 7. **e** 8. **b** 9. **d** 10. **a** 11. **d** 12. **d** 13. **e** 14. **d** 15. **e** 16. **d** 17. **e** 18. **e** 19. **e** 20. **a** 21. **e** 22. **a** 23. **c** 24. **e** 25. **c**

26. Symbols of Chemical Elements
    - Refer to the inside front cover of the textbook, or Table 3.3.

27. Names of Chemical Elements
    - Refer to the inside front cover of the textbook, or Table 3.3.

Chapter 3 Answers to Crossword Puzzle

Across

2. compound
5. sublimation
7. nonmetal
9. state
12. heterogeneous
15. mass
18. metal
20. alloy
21. atom
22. formula
23. ductile

Down

1. semimetal
2. change
3. definite
4. homogeneous
6. malleable
8. kinetic chemical
10. physical
11. potential
13. element
14. symbol
16. substance
17. chemical
19. energy

CHAPTER

4

Models of the Atom

Section 4.1 *Dalton Model of the Atom*

1. Which of Dalton's atomic theory proposals was later proven to be incorrect?
   (a) An element is composed of tiny particles called atoms.
   (b) All atoms of the same element have the same mass.
   (c) Atoms of different elements combine to form compounds.
   (d) Compounds contain atoms in small whole-number ratios.
   (e) Atoms can combine in more than one whole-number ratio.

Section 4.2 *Thomson Model of the Atom*

2. What do the raisins represent in the plum pudding analogy of the atom?
   (a) electrons
   (b) neutrons
   (c) protons
   (d) nuclei
   (e) none of the above

3. Where is the mass of an atom found according to the plum pudding model?
   (a) the mass is found in the electrons
   (b) the mass is found in the protons
   (c) the mass is found in the nucleus
   (d) the mass is distributed homogeneously
   (e) none of the above

Section 4.3 *Rutherford Model of the Atom*

4. What are the approximate diameters of an atom and its nucleus?
   (a) 10^8 cm and 10^{-13} cm, respectively
   (b) 10^{-8} cm and 10^{-13} cm, respectively
   (c) 10^{-13} cm and 10^8 cm, respectively
   (d) 10^{-13} cm and 10^{-8} cm, respectively
   (e) 1 cm and 0.1 cm, respectively

5. State the subatomic particle having a relative charge of zero and an approximate mass of one atomic mass unit.
   (a) alpha
   (b) electron
   (c) neutron
   (d) positron
   (e) proton

Section 4.4 *Atomic Notation*

6. Using atomic notation, indicate the isotope having 10 p^+, 12 n^o, and 10 e^-.
   (a) $^{12}_{10}Ne$
   (b) $^{22}_{10}Ne$
   (c) $^{22}_{12}Ne$
   (d) $^{12}_{10}Mg$
   (e) $^{22}_{12}Mg$

7. How many neutrons are in the nucleus of one atom of oxygen-18?
   (a) 2
   (b) 8
   (c) 10
   (d) 18
   (e) 26

Section 4.5 *Atomic Mass*

8. Nitrogen occurs naturally as ^{14}N and ^{15}N. Which isotope is more abundant? (Hint: Refer to the Periodic Table.)
   (a) nitrogen-7
   (b) nitrogen-8
   (c) nitrogen-14
   (d) nitrogen-15
   (e) none of the above

Element X occurs naturally as X-6 (6.015 amu) and X-7 (7.016 amu). Calculate the atomic mass of X given that the natural abundance of X-7 is 92.5%.

(a) 6.09 amu
(b) 6.50 amu
(c) 6.52 amu
(d) 6.94 amu
(e) 12.5 amu

10. Refer to the periodic table and determine the atomic mass of barium.

(a) 25 amu
(b) 56 amu
(c) 81 amu
(d) 137.33 amu
(e) 193.33 amu

11. Refer to the periodic table and determine the mass number for an important isotope of radon.

(a) 86
(b) 136
(c) 222
(d) 308
(e) none of the above

Section 4.6 *The Wave Nature of Light*

12. Which of the following colors of light has the lowest energy?

(a) blue
(b) red
(c) violet
(d) yellow
(e) all colors have the same energy

13. Which of the following colors of light has the shortest wavelength?

(a) blue
(b) red
(c) violet
(d) yellow
(e) all colors have the same wavelength

14. Which of the following wavelengths of light has the lowest energy?

(a) 440 nm
(b) 470 nm
(c) 540 nm
(d) 650 nm
(e) all wavelengths have the same energy

15. Which of the following wavelengths of light has the lowest frequency?
    (a) 440 nm
    (b) 470 nm
    (c) 540 nm
    (d) 650 nm
    (e) all wavelengths have the same frequency

16. Which of the following wavelengths of light has the fastest velocity?
    (a) 440 nm
    (b) 470 nm
    (c) 540 nm
    (d) 650 nm
    (e) all wavelengths have the same velocity

Section 4.7 *The Quantum Concept*

17. Which of the following instruments gives a quantized measurement?
    (a) 100-mm metric ruler
    (b) 100-g platform balance
    (c) 100-cc hypodermic syringe
    (d) 100-mL volumetric pipet
    (e) none of the above

18. Which of the following is evidence for a quantized change?
    (a) a line spectrum
    (b) a continuous spectrum
    (c) a visible spectrum
    (d) a rainbow
    (e) all of the above

Section 4.8 *Bohr Model of the Atom*

19. Which of these statements is *not* true of the Bohr model of the atom?
    (a) Electrons maintain a constant energy state as they orbit the nucleus.
    (b) Electrons that receive heat energy jump to a higher orbit.
    (c) Electrons gain energy if they drop to an orbit closer to the nucleus.
    (d) Electrons circle the nucleus similar to the way planets circle the sun.
    (e) none of the above

20. Which of the following produces the "atomic fingerprint" for hydrogen?
    (a) atoms of helium reacting
    (b) atoms of helium colliding
    (c) electrons jumping to a higher energy level
    (d) electrons dropping to a lower energy level
    (e) electrons being captured by the nucleus

21. How many photons of light are emitted when the electron in a hydrogen atom drops from energy level 5 to 2?
   (a) 1
   (b) 2
   (c) 3
   (d) 5
   (e) none of the above

Section 4.9 *Energy Levels and Sublevels*

22. How many energy sublevels theoretically exist in the 5th energy level?
   (a) 2
   (b) 5
   (c) 10
   (d) 18
   (e) none of the above

23. What is the maximum number of electrons that can theoretically occupy the 5*d* sublevel?
   (a) 2
   (b) 5
   (c) 6
   (d) 10
   (e) none of the above

24. What is the maximum number of electrons that can occupy the 2nd energy level?
   (a) 2
   (b) 6
   (c) 8
   (d) 10
   (e) 18

Section 4.10 *Electron Configuration*

25. What element has the following electron configuration: $1s^2\ 2s^2\ 2p^6\ 3s^2\ 3p^1$?
   (a) B
   (b) Na
   (c) Al
   (d) K
   (e) Sc

26. What element has the following electron configuration: $1s^2\ 2s^2\ 2p^6\ 3s^2\ 3p^3$?
   (a) N
   (b) Al
   (c) P
   (d) K
   (e) Sc

Section 4.11 *Quantum Mechanical Model of the Atom*

27. Which of the following is a distinction between an orbit and an orbital?
    (a) An orbital represents an energy level.
    (b) An orbital represents a probability boundary.
    (c) An orbital may contain more than one electron.
    (d) An orbital may not gain or lose energy.
    (e) An orbit and an orbital are identical.

28. Which of the following orbitals has a dumbbell shape in the 5th energy level?
    (a) 5*s*
    (b) 4*p*
    (c) 5*p*
    (d) 4*d*
    (e) 5*d*

29. What is the maximum number of electrons that can occupy a 4*d* orbital?
    (a) 2
    (b) 3
    (c) 4
    (d) 6
    (e) 10

30. How many orbitals exist within the 4*d* sublevel?
    (a) 2
    (b) 4
    (c) 5
    (d) 6
    (e) 10

31. What is the maximum number of electrons that can occupy a 4*d* sublevel?
    (a) 2
    (b) 4
    (c) 7
    (d) 10
    (e) 14

32. What is the maximum number of electrons that can occupy the 4th energy level?
    (a) 2
    (b) 4
    (c) 10
    (d) 18
    (e) 32

Chapter 4 Key Terms

Across

2. the distance light travels in one cycle
4. the arrangement of electrons in an atom
8. an atomic model with electron probability (2 words)
14. a model of an atom with electron orbits
16. a spectrum with narrow bands of radiant energy
18. the ______ number refers to the *Z* value
19. atoms that differ only by the number of neutrons
20. an energy level that is designated *s*, *p*, *d*, *f* ...
22. an electron orbit designated 1, 2, 3 ... (2 words)
23. the number of wave cycles of light in one second

Down

1. a positively charged subatomic particle
3. 1/12 the mass of a carbon-12 atom
5. a spectrum showing a broad band of radiant energy
6. a neutral subatomic particle
7. a synonym for a quantum
9. the weighted average mass of isotopes (2 words)
10. the ______ number refers to the *A* value
11. a region of high density in the center of the atom
12. visible radiant energy, violet through red
13. a region of high probability for finding an electron
15. a negatively charged subatomic particle
17. the principle that both the location and energy of an electron cannot be measured precisely
21. the light spectrum from 400–700 nm

Chapter 4 Answers to Self-Test

1. **b** 2. **a** 3. **d** 4. **b** 5. **c** 6. **b** 7. **c** 8. **c** 9. **d** 10. **d** 11. **c** 12. **b** 13. **c** 14. **d** 15. **d** 16. **e** 17. **d** 18. **a** 19. **c** 20. **d** 21. **a** 22. **b** 23. **d** 24. **c** 25. **c** 26. **c** 27. **b** 28. **c** 29. **a** 30. **c** 31. **d** 32. **e**

Chapter 4 Answers to Crossword Puzzle

Across

2. wavelength
4. configuration
8. quantum mechanical
14. Bohr
16. line
18. atomic
19. isotopes
20. sublevel
22. energy level
23. frequency

Down

1. proton
3. amu
5. continuous
6. neutron
7. photon
9. atomic mass
10. mass
11. nucleus
12. light
13. orbital
15. electron
17. uncertainty
21. visible

CHAPTER

5

The Periodic Table

Section 5.1 *Classification of Elements*

1. Which of the following scientists contributed to the development of the periodic table of elements?
   (a) Johann Döbereiner
   (b) Dmitri Mendeleev
   (c) H.G.J. Moseley
   (d) J.A.R. Newlands
   (e) all of the above

2. Why did Mendeleev *not* include germanium (Ge) in his periodic table of 1871?
   (a) Ge is unstable.
   (b) Ge is radioactive.
   (c) Ge is not found naturally.
   (d) Ge was not discovered until 1886.
   (e) none of the above

Section 5.2 *The Periodic Law Concept*

3. The original periodic law was based upon which of the following?
   (a) increasing atomic mass
   (b) increasing atomic number
   (c) increasing isotopic mass
   (d) increasing mass number
   (e) increasing neutron number

4. The modern periodic law is based on which of the following?
   (a) increasing atomic mass
   (b) increasing atomic number
   (c) increasing isotopic mass
   (d) increasing mass number
   (e) increasing neutron number

Section 5.3 *Groups and Periods of Elements*

5. Which fourth period representative element has the lowest atomic number?
   (a) C
   (b) K
   (c) Sc
   (d) Ga
   (e) Y

6. Which fourth period transition element has the lowest atomic number?
   (a) C
   (b) K
   (c) Sc
   (d) Ga
   (e) Y

7. Previous to the 1985 recommendation by IUPAC, what was the designation for the nitrogen family of elements according to the American convention?
   (a) IIIA
   (b) IIIB
   (c) VA
   (d) VB
   (e) none of the above

Section 5.4 *Periodic Trends*

8. Which of the following is a general trend from left to right in the periodic table?
   (a) atomic radius increases; metallic character increases
   (b) atomic radius increases; metallic character decreases
   (c) atomic radius decreases; metallic character increases
   (d) atomic radius decreases; metallic character decreases
   (e) none of the above

9. Predict which of the following elements has the largest atomic radius.
   (a) F
   (b) Cl
   (c) O
   (d) S
   (e) Se

10. Predict which of the following elements has the most nonmetallic character.
   (a) F
   (b) Cl
   (c) O
   (d) S
   (e) Se

Section 5.5 *Properties of Elements*

11. Predict the density of barium, given the density of calcium (1.54 g/cm³) and strontium (2.63 g/cm³).
    (a) 0.45 g/cm³
    (b) 1.09 g/cm³
    (c) 2.09 g/cm³
    (d) 3.72 g/cm³
    (e) 4.17 g/cm³
12. Predict the atomic radius of antimony given the radius of arsenic (0.125 nm) and bismuth (0.155 nm).
    (a) 0.030 nm
    (b) 0.095 nm
    (c) 0.140 nm
    (d) 0.185 nm
    (e) 0.280 nm
13. Predict the boiling point for argon given the boiling points of krypton (–152 °C) and xenon (–107 °C).
    (a) –197 °C
    (b) –130 °C
    (c) –62 °C
    (d) –45 °C
    (e) 45 °C
14. Predict which of the following elements has chemical properties most similar to iron.
    (a) Al
    (b) Ru
    (c) Na
    (d) Ni
    (e) Zn

Section 5.6 *Blocks of Elements*

15. Which energy sublevel is being filled by the elements Ga through Kr?
    (a) 3*d*
    (b) 4*s*
    (c) 4*p*
    (d) 4*d*
    (e) 4*f*
16. Which energy sublevel is being filled by the elements Y through Cd?
    (a) 4*d*
    (b) 5*s*
    (c) 5*p*
    (d) 5*d*
    (e) 5*f*

17. Which of the following is the electron configuration for an atom of tin?
   (a) [Kr] $5s^2\ 4d^{10}\ 4p^2$
   (b) [Kr] $5s^2\ 4d^{10}\ 5p^2$
   (c) [Kr] $5s^2\ 4d^{10}\ 5d^2$
   (d) [Xe] $5s^2\ 4d^{10}\ 5p^2$
   (e) [Xe] $5s^2\ 4d^{10}\ 5d^2$

Section 5.7 *Valence Electrons*

18. Predict the number of valence electrons for a Group IIA/2 element.
   (a) 1
   (b) 2
   (c) 3
   (d) 6
   (e) 8

19. Predict the number of valence electrons for an atom of selenium.
   (a) 2
   (b) 4
   (c) 6
   (d) 16
   (e) 34

Section 5.8 *Electron Dot Formulas*

20. Which of the following is the electron dot formula for an atom of strontium?
   (a) $\mathrm{Sr}\cdot$ (b) $\underset{\cdot}{\mathrm{Sr}}\cdot$ (c) $\cdot\overset{\cdot}{\underset{\cdot}{\mathrm{Sr}}}:$ (d) $\cdot\overset{\cdot}{\underset{\cdot\cdot}{\mathrm{Sr}}}:$ (e) $:\overset{\cdot\cdot}{\underset{\cdot\cdot}{\mathrm{Sr}}}:$

21. Which of the following is the electron dot formula for an atom of oxygen?
   (a) $\mathrm{O}\cdot$ (b) $\underset{\cdot}{\mathrm{O}}\cdot$ (c) $\cdot\overset{\cdot}{\underset{\cdot}{\mathrm{O}}}\cdot$ (d) $\cdot\overset{\cdot}{\underset{\cdot\cdot}{\mathrm{O}}}:$ (e) $:\overset{\cdot}{\underset{\cdot\cdot}{\mathrm{O}}}:$

Section 5.9 *Ionization Energy*

22. Which of the following groups of elements has the highest ionization energy?
   (a) Group IA/1
   (b) Group IIA/2
   (c) Group IIB/12
   (d) Group VIIA/17
   (e) Group VIIIA/18

23. Which of the following elements has the lowest ionization energy?
   (a) Rb
   (b) Sr
   (c) Cs
   (d) Ba
   (e) Xe

. Which of the following is a general trend for the ionization energy of elements in the periodic table?
(a) increases from left to right; increases from bottom to top
(b) increases from left to right; decreases from bottom to top
(c) decreases from left to right; increases from bottom to top
(d) decreases from left to right; decreases from bottom to top
(e) none of the above

Section 5.10 *Ionic Charges*

25. Which of the following groups has a predictable ionic charge of two negative?
(a) Group IIA/2
(b) Group IVB/4
(c) Group IVA/14
(d) Group VIA/16
(e) Group VIIIA/17

26. Which of the following ions is *not* isoelectronic with argon?
(a) S^{2-}
(b) Cl^{-}
(c) K^{+}
(d) Sc^{3+}
(e) V^{4+}

27. What is the core electron configuration for a titanium(II) ion, Ti^{2+}?
(a) [Ar]
(b) [Ar] $4s^2$
(c) [Ar] $3d^2$
(d) [Ar] $4s^2$ $3d^2$
(e) none of the above

28. What is the core electron configuration for a gallium ion, Ga^{3+}?
(a) [Ar]
(b) [Ar] $3d^{10}$
(c) [Ar] $4s^2$ $3d^8$
(d) [Ar] $4s^2$ $3d^{10}$ $4p^1$
(e) none of the above

29. What is the core electron configuration for a fluoride ion, F^{-}?
(a) [Ne]
(b) [Ne] $2p^5$
(c) [Ne] $2p^6$
(d) [Ne] $3s^1$
(e) none of the above

30. What is the core electron configuration for a telluride ion, Te^{2-}?
   (a) [Kr]
   (b) [Kr] $4p^6$
   (c) [Xe]
   (d) [Xe] $5p^6$
   (e) none of the above

Periodic Table Exercise

Refer to a periodic table and identify the element by its chemical symbol that matches each of the following descriptions.

(a) the lowest atomic number alkali metal
(b) the lowest atomic number alkaline earth metal
(c) the lowest atomic number halogen
(d) the lowest atomic number noble gas
(e) the lowest atomic number lanthanide
(f) the lowest atomic number actinide
(g) the lowest atomic number rare earth element
(h) the lowest atomic number transuranium element
(i) the radioactive alkali metal
(j) the radioactive alkaline earth metal
(k) the radioactive lanthanide
(l) the radioactive rare earth element
(m) the radioactive halogen
(n) the radioactive noble gas
(o) the liquid metal at room temperature
(p) the liquid nonmetal at room temperature
(q) the Group IA/1 nonmetal
(r) the Group IIIA/13 semimetal
(s) the Group IVA/14 nonmetal
(t) the Group IB/11 coinage metals

Chapter 5 Key Terms

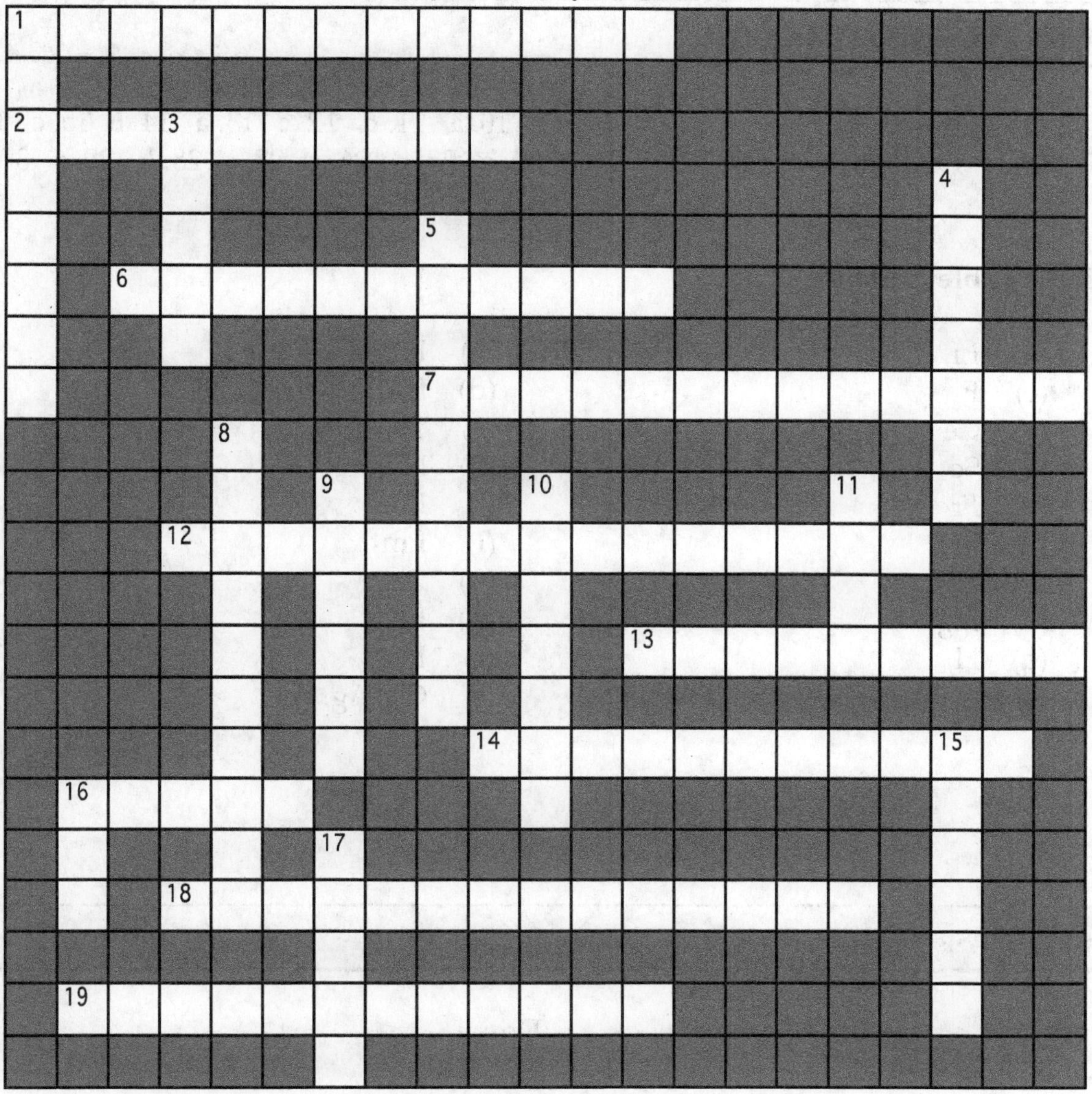

Across

1. the metals Mg, Ca, Sr (2 words)
2. the elements with variable chemical properties
6. Lewis structure (2 words)
7. ions with identical electron configurations
12. the elements filling 4*f* and 5*f* sublevels (2 words)
13. an element such as Sc, Y, La (2 words)
14. properties of elements repeat in a pattern (2 words)
16. charge on an atom after gaining or losing electrons
18. the elements with predictable chemical properties
19. the elements following atomic number 92

Down

1. an element with an atomic number 90–103
3. the unreactive gaseous elements
4. the outer *s* and *p* electrons
5. the process of an atom losing an electron
8. an element with an atomic number 58–71
9. a row of elements
10. an element such as F, Cl, Br, I
11. inner electrons not available for bonding
15. the metals Li, Na, K
16. an atom that has gained or lost electrons
17. a family of elements

Chapter 5 Answers to Self-Test

1. **e** 2. **d** 3. **a** 4. **b** 5. **b** 6. **c** 7. **c** 8. **d** 9. **e** 10. **a** 11. **d** 12. **c** 13. **a** 14. **b** 15. **c** 16. **a** 17. **b** 18. **b** 19. **c** 20. **b** 21. **d** 22. **e** 23. **c** 24. **a** 25. **d** 26. **e** 27. **c** 28. **b** 29. **a** 30. **c**

Periodic Table Exercise

(a) Li
(b) Be
(c) F
(d) He
(e) Ce
(f) Th
(g) Sc
(h) Np
(i) Fr
(j) Ra
(k) Pm
(l) Pm
(m) At
(n) Rn
(o) Hg
(p) Br
(q) H
(r) B
(s) C
(t) Cu, Ag, Au

Chapter 5 Answers to Crossword Puzzle

Across

1. alkaline earth
2. transition
6. electron dot
7. isoelectronic
12. inner transition
13. rare earth
14. periodic law
16. ionic
18. representative
19. transuranium

Down

1. actinide
3. noble
4. valence
5. ionization
8. lanthanide
9. period
10. halogen
11. core
15. alkali
16. ion
17. group

CHAPTER

6

Language of Chemistry

Section 6.1 *Classification of Compounds*

1. The compound $CaSO_4$ is classified as which of the following?
   (a) binary ionic
   (b) ternary ionic
   (c) binary molecular
   (d) binary acid
   (e) ternary oxyacid

2. The compound H_2O is classified as which of the following?
   (a) binary ionic
   (b) ternary ionic
   (c) binary molecular
   (d) binary acid
   (e) ternary oxyacid

3. Aqueous HBr is classified as which of the following?
   (a) binary ionic
   (b) ternary ionic
   (c) binary molecular
   (d) binary acid
   (e) ternary oxyacid

4. The ammonium ion, NH_4^+, is classified as which of the following?
   (a) monoatomic cation
   (b) monoatomic anion
   (c) polyatomic cation
   (d) polyatomic anion
   (e) none of the above

5. The hydroxide ion, OH^-, is classified as which of the following?
   (a) monoatomic cation
   (b) monoatomic anion
   (c) polyatomic cation
   (d) polyatomic anion
   (e) none of the above

Section 6.2 *Monoatomic Ions*

6. What is the systematic IUPAC name for Hg_2^{2+} according to the Stock system?
   (a) mercury ion
   (b) mercury(I) ion
   (c) mercury(II) ion
   (d) mercuric ion
   (e) mercurous ion

7. What is the systematic IUPAC name for Cu^{2+} according to the Latin system?
   (a) copper ion
   (b) copper(II) ion
   (c) cupric ion
   (d) cuprous ion
   (e) cuprum ion

8. What is the systematic IUPAC name for I^-?
   (a) hypoiodite ion
   (b) iodide ion
   (c) iodite ion
   (d) iodate ion
   (e) periodate ion

9. What is the predicted ionic charge of an element in Group VIA/16?
   (a) 2+
   (b) 6+
   (c) 2–
   (d) 6–
   (e) none of the above

Section 6.3 *Polyatomic Ions*

10. What is the systematic IUPAC name for MnO_4^-?
    (a) manganese ion
    (b) manganate ion
    (c) manganite ion
    (d) permanganate ion
    (e) none of the above

11. What is the formula for the hydrogen carbonate ion?
    (a) $C_2H_3O_2^-$
    (b) $C_2H_3O_2^{2-}$
    (c) HCO_3^-
    (d) HCO_3^{2-}
    (e) none of the above

Section 6.4 *Writing Chemical Formulas*

12. What is the chemical formula for the ionic compound composed of calcium ions, Ca^{2+}, and phosphide ions, P^{3-}?
    (a) Ca_2P_2
    (b) Ca_2P_3
    (c) Ca_3P_2
    (d) Ca_3P_3
    (e) Ca_6P_6

13. What is the chemical formula for the ionic compound composed of strontium ions, Sr^{2+} and phosphate ions, PO_4^{3-}?
    (a) $SrPO_4$
    (b) $Sr_2(PO_4)_3$
    (C) $Sr_3(PO_4)_2$
    (d) $Sr_6(PO_4)_6$
    (e) none of the above

14. What is the chemical formula for the ionic compound composed of barium ions, Ba^{2+} and cyanide ions, CN^-?
    (a) BaCN
    (b) $BaCN_2$
    (c) Ba_2CN
    (d) $Ba(CN)_2$
    (e) $Ba_2(CN)_2$

Section 6.5 *Binary Ionic Compounds*

15. What is the ionic charge for the cobalt ion in Co_2S_3?
    (a) 2+
    (b) 2–
    (c) 3+
    (d) 3–
    (e) none of the above

16. What is the systematic IUPAC name for Co_2S_3 according to the Stock system?
    (a) cobalt sulfide
    (b) cobalt(II) sulfide
    (c) cobalt(II) sulfite
    (d) cobalt(III) sulfide
    (e) cobalt(III) sulfite

17. Predict the chemical formula for rubidium fluoride, given the formula of sodium fluoride, NaF.
    (a) RbF
    (b) Rb_2F
    (c) RbF_2
    (d) Rb_2F_3
    (e) Rb_3F_2

Section 6.6 *Ternary Ionic Compounds*

18. What is the ionic charge for the cobalt ion in $Co_3(PO_4)_2$?
    (a) 2+
    (b) 2–
    (c) 3+
    (d) 3–
    (e) none of the above

19. What is the systematic IUPAC name for $Co_3(PO_4)_2$ according to the Latin system?
    (a) cobaltic phosphide
    (b) cobaltous phosphite
    (c) cobaltic phosphite
    (d) cobaltous phosphate
    (e) cobaltic phosphate

20. Predict the chemical formula for aluminum selenide, given the formula of aluminum oxide, Al_2O_3.
    (a) AlSe
    (b) Al_2Se
    (c) $AlSe_3$
    (d) Al_2Se_3
    (e) Al_3Se_2

Section 6.7 *Binary Molecular Compounds*

21. What is the systematic IUPAC name for SF_6?
    (a) sulfur tetrafluoride
    (b) sulfur hexafluoride
    (c) sulfur heptafluoride
    (d) sulfur octafluoride
    (e) sulfur nonafluoride

22. What is the formula for tetraphosphorus trisulfide?
   (a) PS_3
   (b) P_4S
   (c) P_3S_4
   (d) P_4S_3
   (e) P_4S_4

Section 6.8 *Binary Acids*

23. What is the systematic IUPAC name for aqueous HBr?
   (a) hydrogen bromide
   (b) hydrobromous acid
   (c) hydrobromic acid
   (d) bromous acid
   (e) bromic acid

24. What is the formula for hydrosulfuric acid?
   (a) $H_2S(aq)$
   (b) $HSO_3(aq)$
   (c) $HSO_4(aq)$
   (d) $H_2SO_3(aq)$
   (e) $H_2SO_4(aq)$

Section 6.9 *Ternary Oxyacids*

25. What is the systematic IUPAC name for aqueous $HC_2H_3O_2$?
   (a) acetic acid
   (b) carbonic acid
   (c) dicarbonic acid
   (d) hydrogen carbonous acid
   (e) hydrogen carbonic acid

26. What is the formula for sulfuric acid?
   (a) $H_2S(aq)$
   (b) $HSO_3(aq)$
   (c) $HSO_4(aq)$
   (d) $H_2SO_3(aq)$
   (e) $H_2SO_4(aq)$

Chemical Nomenclature Exercise

27. Give an acceptable IUPAC name for each of the following chemical formulas.

(a) KF

(b) CuO

(c) $Zn(NO_3)_2$

(d) $NiSO_4$

(e) $(NH_4)_2CrO_4$

28. Provide the chemical formula for each of the following compounds.

(a) silver iodide

(b) cadmium nitride

(c) manganese(II) chloride

(d) tin(IV) carbonate

(e) iron(II) hydroxide

Chapter 6 Key Terms

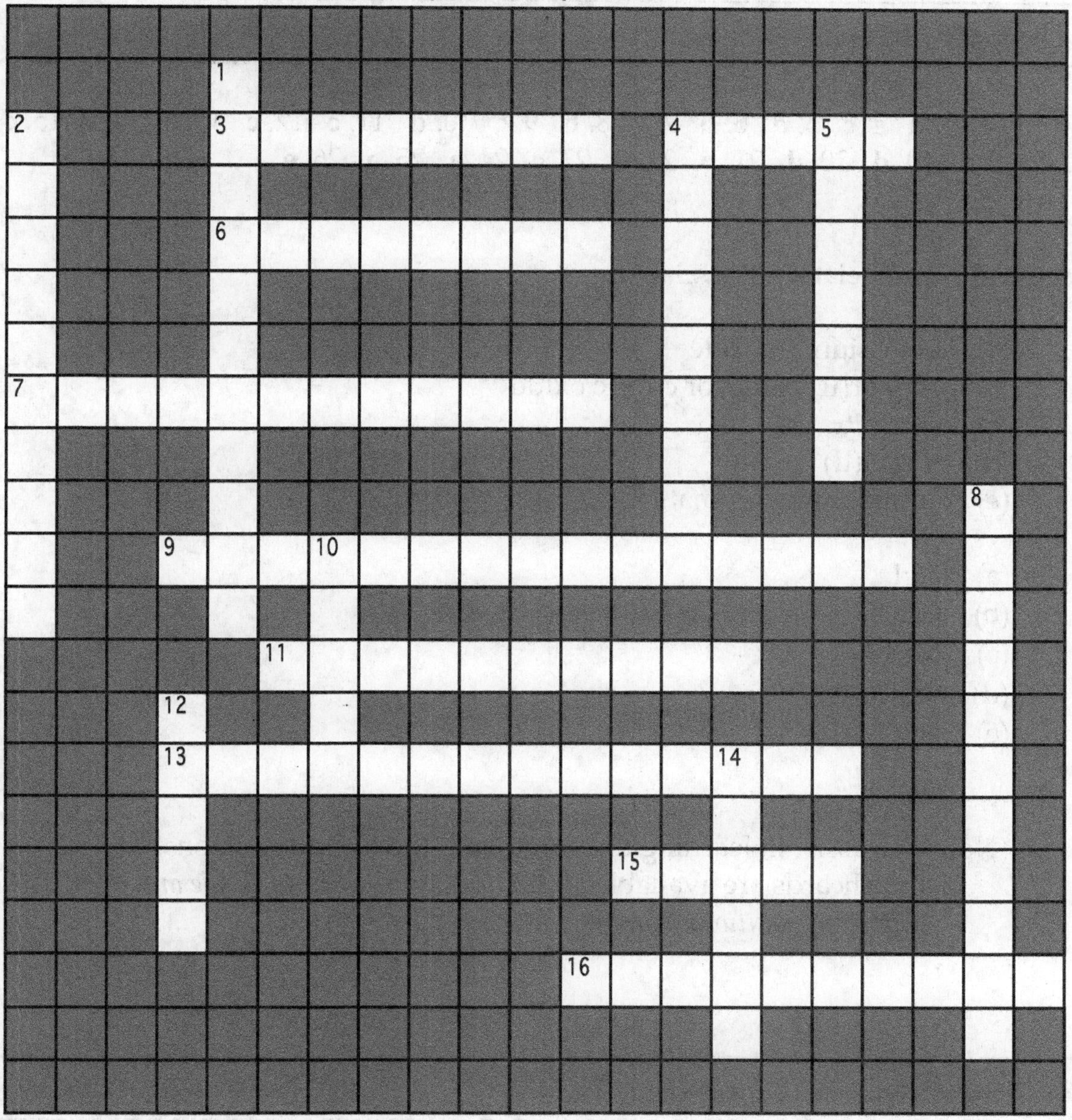

Across

3. a polyatomic anion containing oxygen
4. International Union of Pure and Applied Chemistry
6. the simplest particle in a molecular compound
7. a compound of a metal and two nonmetals (2 words)
9. a compound of two nonmetals (2 words)
11. classification of aqueous HCl (2 words)
13. classification of aqueous H_2SO_4 (2 words)
15. –ous and –ic suffix naming system
16. a single-atom ion

Down

1. a particle in an ionic compound (2 words)
2. a multiple-atom ion
4. a substance not containing carbon
5. a solution of a substance dissolved in water
8. a compound of a metal and a nonmetal (2 words)
10. any negatively charged ion
12. Roman numeral naming system
14. any positively charged ion

Chapter 6 Answers to Self-Test

1. **b** 2. **c** 3. **d** 4. **c** 5. **d** 6. **b** 7. **c** 8. **b** 9. **c** 10. **d** 11. **c** 12. **c** 13. **c** 14. **d** 15. **c** 16. **d** 17. **a** 18. **a** 19. **d** 20. **d** 21. **b** 22. **d** 23. **c** 24. **a** 25. **a** 26. **e**

Chemical Nomenclature Exercise

27.
(a) potassium fluoride
(b) copper(II) oxide, or cupric oxide
(c) zinc nitrate
(d) nickel(II) sulfate
(e) ammonium chromate

28.
(a) AgI
(b) Cd_3N_2
(c) $MnCl_2$
(d) $Sn(CO_3)_2$
(e) $Fe(OH)_2$

(Note: To assist in learning the names and the formulas of ions, flashcards are available at the back of this *Study Guide and Selected Solutions Manual.*)

Chapter 6 Answers to Crossword Puzzle

Across
3. oxyanion
4. IUPAC
6. molecule
7. ternary ionic
9. binary molecular
11. binary acid
13. ternary oxyacid
15. Latin
16. monoatomic

Down
1. formula unit
2. polyatomic
4. inorganic
5. aqueous
8. binary ionic
10. anion
12. Stock
14. cation

CHAPTER

7

Chemical Reactions

Section 7.1 *Evidence for Chemical Reactions*

1. Which of the following is evidence for a chemical reaction?
   (a) a gas is detected
   (b) a precipitate is formed
   (c) a color change is observed
   (d) an energy change is noted
   (e) all of the above

2. Which of the following is *not* evidence for a chemical reaction producing a gas?
   (a) an odor is detected
   (b) a color change is observed
   (c) a flaming splint is extinguished
   (d) a glowing splint bursts into flames
   (e) bubbles are observed

Section 7.2 *Writing Chemical Equations*

3. Which of the statements below best describes the following reaction?

$$2\ AgClO_3(s) \xrightarrow{\Delta} 2\ AgCl(s) + 3\ O_2(g)$$

   (a) Silver chlorate gives silver chloride and oxygen.
   (b) Silver chlorate is heated to give silver chloride and oxygen.
   (c) Silver chlorate gives solid silver chloride and oxygen gas.
   (d) Silver chlorate is heated to give solid silver chloride and oxygen gas.
   (e) Solid silver chlorate is heated to give solid silver chloride and oxygen gas.

Section 7.3 *Balancing Chemical Equations*

4. Which of the following is *not* a general direction for balancing an equation?
   (a) write correct formulas for reactants and products
   (b) begin balancing with the most complex formula
   (c) balance polyatomic ions as a single unit
   (d) balance ionic compounds as a single unit
   (e) check each reactant and product to verify the coefficients

5. What is the coefficient of carbon dioxide after balancing the following equation?

$$\_CuHCO_3(s) \xrightarrow{\Delta} \_Cu_2CO_3(s) + \_H_2O(g) + \_CO_2(g)$$

(a) 1 (b) 2 (c) 3 (d) 4 (e) 6

6. What is the coefficient of water after balancing the following equation?

$$\_HNO_3(aq) + \_Co(OH)_2(aq) \rightarrow \_Co(NO_3)_2(aq) + \_H_2O(l)$$

(a) 1 (b) 2 (c) 3 (d) 4 (e) 6

7. What is the coefficient of gold(III) sulfate after balancing the following equation?

$$\_ZnSO_4(aq) + \_Au(NO_3)_3(aq) \rightarrow \_Au_2(SO_4)_3(s) + \_Zn(NO_3)_2(aq)$$

(a) 1 (b) 2 (c) 3 (d) 4 (e) 6

8. What is the coefficient of lead metal after balancing the following equation?

$$\_Al(s) + \_Pb(C_2H_3O_2)_2(aq) \rightarrow \_Al(C_2H_3O_2)_3(aq) + \_Pb(s)$$

(a) 1 (b) 2 (c) 3 (d) 4 (e) 6

Section 7.4 *Classifying Chemical Reactions*

9. What type of chemical reaction is illustrated in Exercise 8?
   (a) combination reaction
   (b) decomposition reaction
   (c) single-replacement reaction
   (d) double-replacement reaction
   (e) neutralization reaction

ection 7.5 *Combination Reactions*

10. What is the product predicted from the following combination reaction?

$$Li(s) \quad + \quad O_2(g) \quad \xrightarrow{\Delta}$$

(a) LiO
(b) Li_2O
(c) LiO_2
(d) Li_2O_3
(e) Li_3O_2

11. What is the product predicted from the following combination reaction?

$$S(s) \quad + \quad O_2(g) \quad \xrightarrow{\Delta}$$

(a) SO
(b) S_2O
(c) SO_2
(d) SO_3
(e) It is impossible to predict a single product for a nonmetal and oxygen without more information.

12. What is the product predicted from the following combination reaction?

$$Ni(s) \quad + \quad Br_2(g) \quad \xrightarrow{\Delta}$$

(a) NiBr
(b) Ni_2Br
(c) $NiBr_2$
(d) Ni_2Br_3
(e) Ni_3Br_2

Section 7.6 *Decomposition Reactions*

13. What are the products from the following decomposition reaction?

$$Bi(HCO_3)_3(s) \quad \xrightarrow{\Delta}$$

(a) Bi, H_2O, and CO_2
(b) Bi_2CO_3, H_2, and CO_2
(c) $Bi_2(CO_3)_3$ and H_2O
(d) Bi_2CO_3, H_2O, and CO_2
(e) $Bi_2(CO_3)_3$, H_2O, and CO_2

14. What are the products from the following decomposition reaction?

$$Co_2(CO_3)_3(s) \xrightarrow{\Delta}$$

(a) Co and CO_2
(b) CoO and CO_2
(c) Co_2O_3 and CO_2
(d) CoO, H_2O, and CO_2
(e) Co_2O_3, H_2O, and CO_2

15. What are the products from the following decomposition reaction?

$$Ca(ClO_3)_2(s) \xrightarrow{\Delta}$$

(a) Ca and CO_2
(b) Ca, Cl_2, and O_2
(c) $CaCl_2$ and H_2O
(d) $CaCl_2$ and O_2
(e) $CaCl_2$ and CO_2

Section 7.7 *The Activity Series Concept*

16. Which of the following metals reacts with aqueous $FeSO_4$?
Partial Activity Series: Mg > Al > Zn > Fe > (H)
(a) Al
(b) Mg
(c) Zn
(d) all of the above
(e) none of the above

17. Which of the following metals reacts with aqueous $AgNO_3$?
Partial Activity Series: (H) > Cu > Ag > Hg > Au
(a) Au
(b) Cu
(c) Hg
(d) all of the above
(e) none of the above

18. Which of the following metals does *not* react with sulfuric acid?
Partial Activity Series: Mn > Cd > Co > (H) > Ag
(a) Ag
(b) Cd
(c) Co
(d) Mn
(e) Sn

19. Which of the following metals reacts with water at 25 °C?
(a) Al
(b) Fe
(c) Zn
(d) Mg
(e) Sr

Section 7.8 *Single-Replacement Reactions*

20. What are the products from the following single-replacement reaction?

$$Mg(s) + AgNO_3(aq) \rightarrow$$

(a) Ag and $Mg(NO_3)_2$
(b) Ag and $Mg(NO_2)_2$
(c) Ag_2O and $Mg(NO_3)_2$
(d) Ag_2O and $Mg(NO_2)_2$
(e) no reaction

21. What are the products from the following single-replacement reaction?

$$Zn(s) + H_2SO_4(aq) \rightarrow$$

(a) ZnO and H_2SO_3
(b) ZnO and H_2S
(c) $ZnSO_4$ and H_2
(d) $ZnSO_4$ and H_2O
(e) no reaction

22. What are the products from the following single-replacement reaction?

$$Na(s) + H_2O(l) \rightarrow$$

(a) Na_2O and H_2
(b) Na_2O and H_2O
(c) NaOH and H_2
(d) NaOH and H_2O
(e) no reaction

Section 7.9 *Solubility Rules*

23. Which of the following solid compounds is insoluble in water? (Refer to the *Solubility Rules* in Appendix D of the textbook.)
    (a) $(NH_4)_2CO_3$
    (b) K_2CrO_4
    (c) $PbSO_4$
    (d) Na_2S
    (e) $Sr(OH)_2$

Section 7.10 *Double-Replacement Reactions*

24. What are the products from the following double-replacement reaction?

$$BaCl_2(aq) \quad + \quad K_2SO_4(aq) \quad \rightarrow$$

    (a) BaS and $KClO_4$
    (b) $BaSO_3$ and KCl
    (c) $BaSO_3$ and $KClO_4$
    (d) $BaSO_4$ and KCl
    (e) $BaSO_4$ and $KClO_4$

Section 7.11 *Neutralization Reactions*

25. What are the products from the following neutralization reaction?

$$HNO_3(aq) \quad + \quad NH_4OH(aq) \quad \rightarrow$$

    (a) NH_3NO_3 and H_2O
    (b) NH_3NO_3 and O_2
    (c) NH_4NO_3 and H_2O
    (d) NH_4NO_3 and O_2
    (e) NH_4NO_2 and H_2O

Writing Chemical Equations for Chemical Reactions

26. Write a balanced chemical equation for the reaction of aqueous strontium nitrate with aqueous sodium sulfate to yield a precipitate of strontium sulfate and aqueous sodium nitrate.

27. Write a balanced chemical equation for the reaction of hydrochloric acid with aqueous barium hydroxide to give aqueous barium chloride and water.

Chapter 7 Key Terms

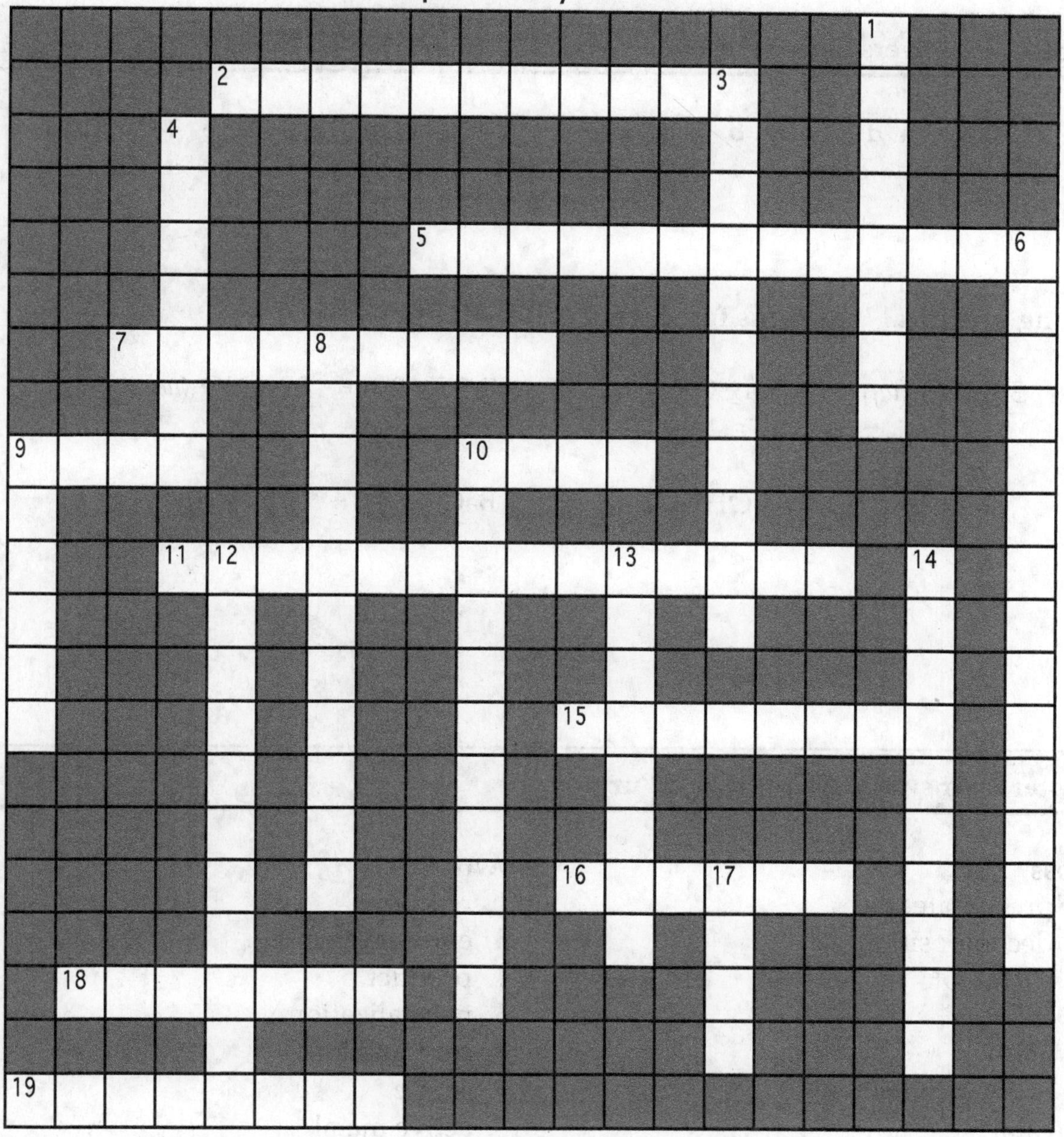

Across

2. an insoluble salt in solution
5. a type of reaction involving a single reactant
7. a number in a chemical formula
9. a compound produced from an acid–base reaction
10. releases hydrogen ions in solution
11. a relative order of metals to react (2 words)
15. a substance that speeds up a chemical reaction
16. a type of replacement reaction that switches ions
18. a solution with a substance dissolved in water
19. a reaction requires this substance

Down

1. a chemical change
3. a reaction that absorbs heat
4. a reaction yields this substance
6. a type of reaction involving an acid and base
8. a type of reaction involving two elements
9. a type of replacement reaction of a metal and acid
10. an element that reacts with water (2 words)
12. a number preceding a formula in an equation
13. describing a reaction using formulas and symbols
14. a reaction that releases heat
17. releases hydroxide ions in solution

Chapter 7 Answers to Self-Test

1. **e** 2. **b** 3. **e** 4. **d** 5. **a** 6. **b** 7. **a** 8. **c** 9. **c** 10. **b** 11. **e** 12. **c** 13. **e** 14. **c** 15. **d** 16. **d** 17. **b** 18. **a** 19. **e** 20. **a** 21. **c** 22. **c** 23. **c** 24. **d** 25. **c**

Writing Chemical Equations for Chemical Reactions

26. $Sr(NO_3)_2(aq) + Na_2SO_4(aq) \rightarrow SrSO_4(s) + 2\ NaNO_3(aq)$

27. $2\ HCl(aq) + Ba(OH)_2(aq) \rightarrow BaCl_2(aq) + 2\ HOH(l)$

Chapter 7 Answers to Crossword Puzzle

Across

2. precipitate
5. decomposition
7. subscript
9. salt
10. acid
11. activity series
15. catalyst
16. double
18. aqueous
19. reactant

Down

1. reaction
3. endothermic
4. product
6. neutralization
8. combination
9. single
10. active metal
12. coefficient
13. equation
14. exothermic
17. base

CHAPTER

8

The Mole Concept

Section 8.1 *Avogadro's Number*

1. How many atoms of carbon have a mass equal to its atomic mass, 12.01 g?
   (a) 1
   (b) 6
   (c) 12
   (d) 12.01
   (e) 6.02×10^{23}

2. What is the mass of Avogadro's number of iron atoms?
   (a) 55.85 amu
   (b) 55.85 grams
   (c) 6.02×10^{23} amu
   (d) 6.02×10^{23} grams
   (e) 55.85×10^{23} grams

Section 8.2 *Mole Calculations I*

3. Which of the following is equal to 1 mol of substance?
   (a) 6.02×10^{23} sodium atoms, Na
   (b) 6.02×10^{23} chlorine molecules, Cl_2
   (c) 6.02×10^{23} sodium chloride formula units, NaCl
   (d) 6.02×10^{23} chloride ions, Cl^-
   (e) all of the above

4. How many xenon atoms are in 0.250 mol of the noble gas?
   (a) 1.51×10^{22} atoms
   (b) 1.51×10^{23} atoms
   (c) 1.51×10^{24} atoms
   (d) 2.41×10^{23} atoms
   (e) 2.41×10^{24} atoms

5. How many moles of argon correspond to 7.52×10^{22} atoms of the inert gas?
   (a) 0.0125 mol
   (b) 0.0801 mol
   (c) 0.125 mol
   (d) 0.801 mol
   (e) 1.25 mol

Section 8.3 *Molar Mass*

6. What is the approximate molar mass of trinitrotoluene (TNT), $C_7H_5(NO_2)_3$?
   (a) 63.07 g/mol
   (b) 135.13 g/mol
   (c) 189.21 g/mol
   (d) 227.15 g/mol
   (e) 405.39 g/mol

Section 8.4 *Mole Calculations II*

7. What is the mass of 4.50×10^{22} atoms of gold?
   (a) 0.0679 g
   (b) 0.0748 g
   (c) 13.3 g
   (d) 14.7 g
   (e) 197 g

8. How many methane molecules, CH_4, have a mass of 3.20 g?
   (a) 1.20×10^{23} molecules
   (b) 1.20×10^{24} molecules
   (c) 1.93×10^{24} molecules
   (d) 3.01×10^{23} molecules
   (e) 3.01×10^{24} molecules

9. What is the mass of a formula unit of cuprous sulfide, Cu_2S?
   (a) 1.04×10^{-26} g
   (b) 1.66×10^{-24} g
   (c) 2.64×10^{-22} g
   (d) 3.79×10^{21} g
   (e) 9.57×10^{25} g

Section 8.5 *Molar Volume*

10. One mole of which of the following gases occupies 22.4 L at STP?
    (a) ammonia, NH_3
    (b) helium, He
    (c) hydrogen, H_2
    (d) oxygen, O_2
    (e) all of the above

11. What is the density of nitrogen gas, N_2, at STP?
    (a) 0.625 g/L
    (b) 0.800 g/L
    (c) 1.25 g/L
    (d) 1.60 g/L
    (e) 22.4 g/L

12. If the density of nitrous oxide is 1.96 g/L at STP, what is the molar mass of laughing gas?
    (a) 11.4 g/mol
    (b) 19.6 g/mol
    (c) 22.4 g/mol
    (d) 43.9 g/mol
    (e) 196 g/mol

Section 8.6 *Mole Calculations III*

13. What is the volume of 1.51×10^{24} molecules of ammonia gas, NH_3, at STP?
    (a) 0.112 L
    (b) 0.399 L
    (c) 2.51 L
    (d) 8.93 L
    (e) 56.2 L

14. What is the mass of 0.500 liter of oxygen gas, O_2, at STP?
    (a) 0.286 g
    (b) 0.714 g
    (c) 3.50 g
    (d) 6.40 g
    (e) 112 g

Section 8.7 *Percent Composition*

15. The formula for the illegal drug cocaine is $C_{17}H_{21}NO_4$. What is the percentage of nitrogen in the compound?
    (a) 4.62%
    (b) 6.99%
    (c) 21.1%
    (d) 32.6%
    (e) 67.3%

16. Phenacyl chloride is used as tear gas for riot control and has the formula C_7H_7OCl. What is the percentage of carbon in the compound?
    (a) 4.96%
    (b) 8.42%
    (c) 11.2%
    (d) 24.9%
    (e) 59.0%

Section 8.8 *Empirical Formula*

17. If 0.250 mol V reacts with 0.375 mol O, what is the empirical formula of vanadium oxide?
    (a) VO
    (b) V_2O_3
    (c) V_2O_5
    (d) V_3O_2
    (e) V_5O_2

18. If 1.500 g of vanadium metal react with oxygen gas to give 2.679 g of vanadium oxide, what is the empirical formula of the product?
    (a) VO
    (b) V_2O_3
    (c) V_2O_5
    (d) V_3O_2
    (e) V_5O_2

19. Acetaldehyde is used in the manufacture of perfumes, flavors, and dyes. Calculate the empirical formula for acetaldehyde given that its percent composition is: 54.53% C, 9.15% H, and 36.32% O.
    (a) $C_1H_1O_1$
    (b) $C_2H_4O_1$
    (c) $C_2H_3O_1$
    (d) $C_5H_9O_2$
    (e) $C_6H_9O_3$

Section 8.9 *Molecular Formula*

20. Find the molecular formula for aspirin given that its empirical formula is $C_9H_8O_4$ and its approximate molar mass is 175 g/mol.
    (a) $C_1H_1O_1$
    (b) $C_2H_2O_1$
    (c) $C_9H_8O_4$
    (d) $C_{18}H_{16}O_8$
    (e) $C_{27}H_{24}O_{12}$

21. Lindane is an insecticide that causes dizziness and diarrhea in humans. Find the molecular formula for lindane given that its percent composition is: 24.8% C, 2.1% H, and 73.1% Cl. The approximate molar mass of lindane is 290 g/mol.
    (a) $C_1H_1Cl_1$
    (b) $C_3H_3Cl_3$
    (c) $C_1H_1Cl_6$
    (d) $C_6H_6Cl_6$
    (e) $C_{12}H_1Cl_{35}$

Analogies for Avogadro's Number

22. Given 6.02×10^{23} hydrogen atoms (0.07 nm in diameter), placed side-by-side, which of the following is the best estimate of the total length?
    (a) diameter of a pin
    (b) centimeter
    (c) meter stick
    (d) kilometer
    (e) million times Earth's circumference

23. Given 6.02×10^{23} steel 16-pound shotput balls, which of the following is the best estimate of their total mass?
    (a) microgram
    (b) kilogram
    (c) automobile
    (d) Earth
    (e) Universe

24. Given 6.02×10^{23} softballs, which of the following is the best estimate of their total volume?
    (a) 1000-mL beaker
    (b) Dallas Cowboys Stadium
    (c) Grand Canyon
    (d) Earth
    (e) Universe

Empirical and Molecular Formulas

25. The chemical name of milk sugar is lactose, and is comprised of the simpler sugars glucose and galactose. If the percent composition of galactose is 40.00% C, 6.72% H, 53.29% O, what is its empirical formula?

26. If the approximate molar mass of galactose is about 180 g/mol, what is the molecular formula of galactose?

Chapter 8 Key Terms

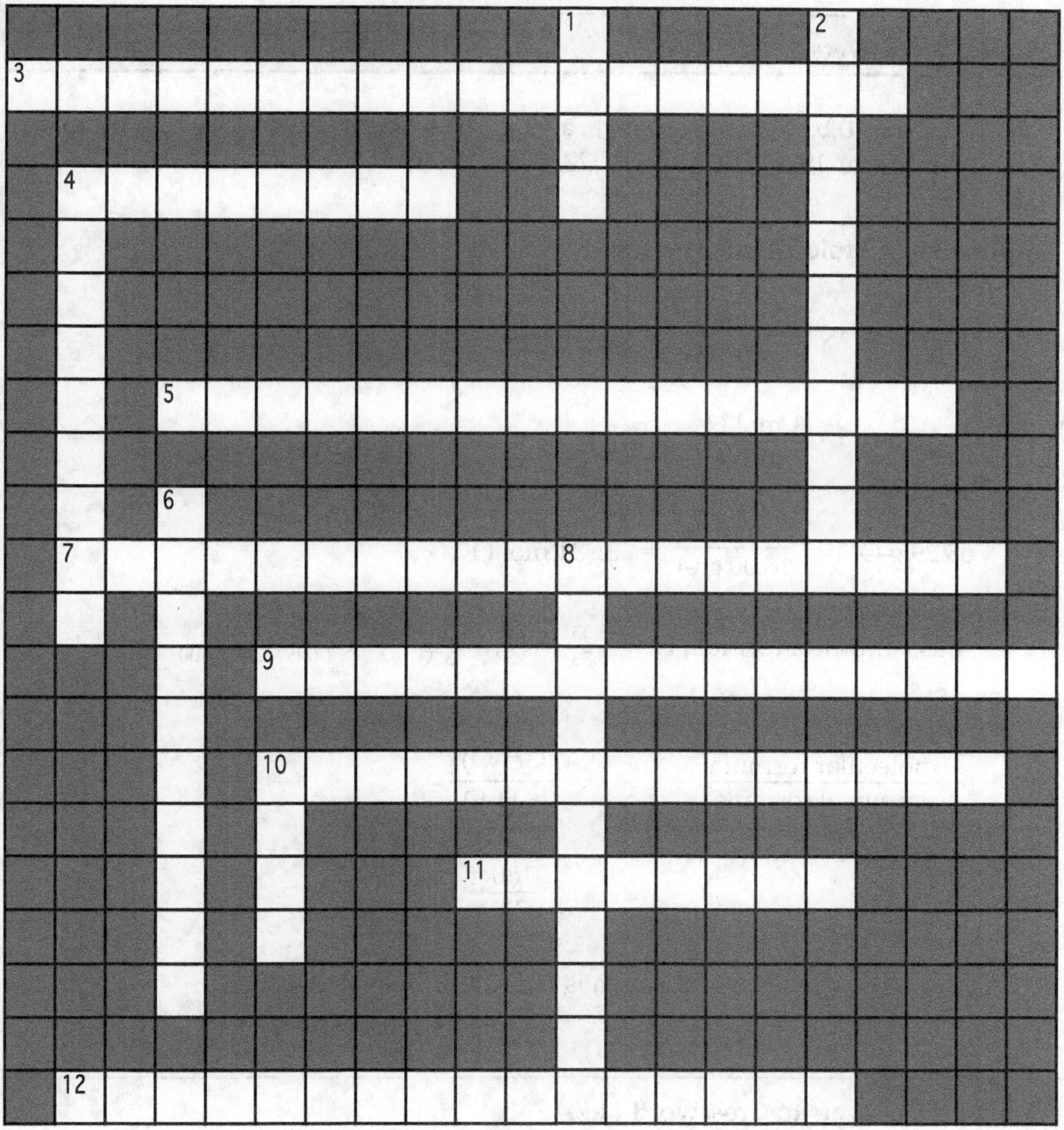

Across

3. the amount of each element in a compound (2 words)
4. particle in a molecular compound
5. the simplest ratio of atoms in a molecule (2 words)
7. 0 °C (2 words)
9. the actual ratio of atoms in a molecule (2 words)
10. number of atoms in 12.01 g of carbon (2 words)
11. stated equal volumes of gas have equal molecules
12. 1 atm (2 words)

Down

1. amount of substance with 6.02×10^{23} particles
2. particle in an ionic compound (2 words)
4. the mass of 1 mol of substance (2 words)
6. usually expressed in grams per liter (2 words)
8. the space occupied by 1 mol of gas (2 words)
10. a representative particle of an element

Chapter 8 Answers to Self-Test

1. e 2. b 3. e 4. b 5. c 6. d 7. d 8. a 9. c 10. e 11. c 12. d 13. e 14. b 15. a 16. e 17. b 18. c 19. b 20. c 21. d 22. e 23. d 24. d

Empirical and Molecular Formulas

25. $$40.00\ \cancel{\text{g C}} \times \frac{1\ \text{mol C}}{12.01\ \cancel{\text{g C}}} = 3.331\ \text{mol C}$$

$$6.72\ \cancel{\text{g H}} \times \frac{1\ \text{mol H}}{1.01\ \cancel{\text{g H}}} = 6.65\ \text{mol H}$$

$$53.29\ \cancel{\text{g O}} \times \frac{1\ \text{mol O}}{16.00\ \cancel{\text{g O}}} = 3.331\ \text{mol O}$$

thus, empirical formula is $C_{\frac{3.33}{3.33}} H_{\frac{6.65}{3.33}} O_{\frac{3.33}{3.33}} = C_{1.00}H_{1.99}O_{1.00} = CH_2O$

26. $$\frac{\text{molecular formula}}{\text{empirical formula}} = \frac{(CH_2O)_n}{CH_2O}$$

$$n = \frac{180\ \text{g}}{30\ \text{g}} = 6$$

thus, molecular formula is $(CH_2O)_6 = C_6H_{12}O_6$

Chapter 8 Answers to Crossword Puzzle

Across

3. percent composition
4. molecule
5. empirical formula
7. standard temperature
9. molecular formula
10. Avogadro's number
11. Avogadro
12. standard pressure

Down

1. mole
2. formula unit
4. molar mass
6. gas density
8. molar volume
10. atom

Chemical Equation Calculations

CHAPTER 9

Section 9.1 *Interpreting a Chemical Equation*

1. How many moles of oxygen gas, O_2, react with 2 moles of nitrogen monoxide gas, NO, according to the following equation?

$$\_NO(g) + \_O_2(g) \xrightarrow{\Delta/UV} \_NO_2(g)$$

(a) 1 mol
(b) 2 mol
(c) 3 mol
(d) 4 mol
(e) none of the above

2. How many molar masses of nitrogen dioxide gas are produced from 2 molar masses of nitrogen monoxide gas according to the following equation?

$$\_NO(g) + \_O_2(g) \xrightarrow{\Delta/UV} \_NO_2(g)$$

(a) 1 MM
(b) 2 MM
(c) 3 MM
(d) 4 MM
(e) none of the above

3. How many liters of oxygen gas react to produce 2 L of nitrogen dioxide gas (at similar conditions) according to the following equation?

$$\_NO(g) + \_O_2(g) \xrightarrow{\Delta/UV} \_NO_2(g)$$

(a) 1 L
(b) 2 L
(c) 3 L
(d) 4 L
(e) none of the above

4. In an experiment 0.300 g of NO gas reacts to produce 0.460 g of NO_2 gas. Using the conservation of mass law, predict the mass of reacting oxygen gas.

$$\_NO(g) + \_O_2(g) \xrightarrow{\Delta/UV} \_NO_2(g)$$

(a) 0.0800 g
(b) 0.160 g
(c) 0.320 g
(d) 0.380 g
(e) 0.760 g

Section 9.2 *Mole–Mole Relationships*

5. How many moles of water react with 0.500 mol of lithium metal according to the following reaction?

$$\_Li(s) + \_H_2O(l) \rightarrow \_LiOH(aq) + \_H_2(g)$$

(a) 0.125 mol
(b) 0.250 mol
(c) 0.500 mol
(d) 1.00 mol
(e) 2.00 mol

6. How many moles of hydrogen gas are produced from the reaction of 0.500 mol of sodium metal?

$$\_Na(s) + \_H_2O(l) \rightarrow \_NaOH(aq) + \_H_2(g)$$

(a) 0.125 mol
(b) 0.250 mol
(c) 0.500 mol
(d) 1.00 mol
(e) 2.00 mol

7. How many moles of water must react in order to produce 0.500 mol of potassium hydroxide?

$$\_K(s) + \_H_2O(l) \rightarrow \_KOH(aq) + \_H_2(g)$$

(a) 0.125 mol
(b) 0.250 mol
(c) 0.500 mol
(d) 1.00 mol
(e) 2.00 mol

Section 9.3 *Types of Stoichiometry Problems*

8. Classify the following type of stoichiometry problem: *How many grams of sulfur must react with an excess volume of oxygen in order to yield 0.500 L of sulfur dioxide gas?*
   (a) mass–mass problem
   (b) mass–volume problem
   (c) volume–volume problem
   (d) mole–mole
   (e) none of the above

9. Classify the following type of stoichiometry problem: *How many grams of sulfur must react with an excess volume of oxygen in order to yield 0.500 g of sulfur dioxide gas?*
   (a) mass–mass problem
   (b) mass–volume problem
   (c) volume–volume problem
   (d) mole–mole
   (e) none of the above

10. Classify the following type of stoichiometry problem: *How many liters of sulfur must react with an excess volume of oxygen in order to yield 50.0 cm^3 of sulfur dioxide gas?*
   (a) mass–mass problem
   (b) mass–volume problem
   (c) volume–volume problem
   (d) mole–mole
   (e) none of the above

11. In general, how many unit factors are necessary to solve mass–mass stoichiometry problems?
   (a) one
   (b) two
   (c) three
   (d) four
   (e) five

12. In general, how many unit factors are necessary to solve mass–volume stoichiometry problems?
   (a) one
   (b) two
   (c) three
   (d) four
   (e) five

13. Which of the following steps is *not* necessary to solve a volume–volume stoichiometry problem?
   (a) write a balanced chemical equation for the reaction
   (b) calculate the moles of known gas given its volume
   (c) convert the volume of known gas to the volume of unknown gas
   (d) determine the volume of known gas
   (e) determine the volume of unknown gas

Section 9.4 *Mass–Mass Problems*

14. How many grams of iron are produced from the reaction of 500.0 g of aluminum metal?

$$\_FeO(l) + \_Al(l) \xrightarrow{\Delta} \_Fe(l) + \_Al_2O_3(s)$$

(a) 345 g
(b) 689 g
(c) 1030 g
(d) 1550 g
(e) 3100 g

15. How many grams of aluminum metal must react to give 500.0 g of iron?

$$\_FeO(l) + \_Al(l) \xrightarrow{\Delta} \_Fe(l) + \_Al_2O_3(s)$$

(a) 80.6 g
(b) 161 g
(c) 242 g
(d) 362 g
(e) 483 g

Section 9.5 *Mass–Volume Problems*

16. What STP volume of oxygen is released from the decomposition of 2.166 g of mercuric oxide (216.59 g/mol)?

$$\_HgO(s) \xrightarrow{\Delta} \_Hg(l) + \_O_2(g)$$

(a) 0.112 L
(b) 0.224 L
(c) 0.448 L
(d) 1.12 L
(e) 2.24 L

17. How many grams of mercuric oxide (216.59 g/mol) must decompose to release 0.375 L of oxygen gas at STP?

$$\_HgO(s) \xrightarrow{\Delta} \_Hg(l) + \_O_2(g)$$

(a) 0.0752 g
(b) 1.82 g
(c) 3.63 g
(d) 7.25 g
(e) 14.5 g

ection 9.6 *Volume–Volume Problems*

18. Calculate the volume of oxygen gas that reacts with 32.0 mL of sulfur dioxide. (Assume constant conditions.)

$$\_SO_2(g) + \_O_2(g) \xrightarrow{650\ °C} \_SO_3(g)$$

(a) 8.00 mL
(b) 16.0 mL
(c) 32.0 mL
(d) 64.0 mL
(e) 96.0 mL

19. Calculate the volume of nitrogen gas that produces 50.0 mL of ammonia, NH_3. (Assume constant conditions.)

$$\_N_2(g) + \_H_2(g) \xrightarrow{500\ °C/500\ atm} \_NH_3(g)$$

(a) 6.25 mL
(b) 12.5 mL
(c) 25.0 mL
(d) 50.0 mL
(e) 100.0 mL

ection 9.7 *Limiting Reactant Concept*

20. If 1.00 mol of chromium reacts with 1.00 mol of oxygen gas according to the following equation, how many moles of chromium(III) oxide are produced?

$$\_Cr(s) + \_O_2(g) \xrightarrow{\Delta} \_Cr_2O_3(s)$$

(a) 0.500 mol
(b) 0.667 mol
(c) 1.00 mol
(d) 1.50 mol
(e) 2.00 mol

Section 9.8 *Limiting Reactant Problems*

21. If 20.0 mL of methane gas reacts with 20.0 mL of oxygen gas, what is the volume of carbon dioxide gas produced (assume constant conditions)?

$$\_CH_4(g) + \_O_2(g) \xrightarrow{spark} \_CO_2(g) + \_H_2O(g)$$

(a) 5.00 mL
(b) 10.0 mL
(c) 20.0 mL
(d) 40.0 mL
(e) 80.0 mL

22. If 10.0 g of aluminum metal react with 10.0 g of sulfur according to the following equation, how many grams of aluminum sulfide are produced?

$$\_Al(s) \quad + \quad \_S(s) \quad \xrightarrow{\Delta} \quad \_Al_2S_3(s)$$

(a) 15.6 g
(b) 20.0 g
(c) 27.8 g
(d) 46.8 g
(e) 55.7 g

Section 9.9 *Percent Yield*

23. Starting with 0.657 g of lead(II) nitrate, a student collects 0.905 g of precipitate. If the calculated mass of precipitate is 0.914 g, what is the percent yield?
(a) 71.9%
(b) 72.6%
(c) 99.0%
(d) 101%
(e) 138%

Stoichiometry Problems – Show all work for the following.

24. What is the mass of solid silver chloride produced from the reaction of excess aqueous silver nitrate and 0.333 g of barium chloride?

25. What is the volume of oxygen gas liberated at STP when 1.555 g of solid silver chlorate is decomposed with heat to give solid silver chloride?

Chapter 9 Key Terms

Across

1. liters-to-liters type of problem (2 words)
3. grams-to-liters type of problem (2 words)
4. the experimental yield from a reaction
9. element produced from the blast furnace process
15. Lavoisier's law (3 words)
16. the substance produced from the Haber process
17. ratio of moles of reactants and products (2 words)
18. 0 °C and 1 atm

Down

2. a single particle composed of nonmetal atoms
5. Gay-Lussac's law (2 words)
6. relates quantities in a balanced equation
7. controls maximum amount of product (2 words)
8. the calculated yield from a reaction
10. ratio of actual to theoretical yield (2 words)
11. stated equal volumes—same number of molecules
12. the mass of one mole of substance (2 words)
13. the volume of one mole of gas (2 words)
14. grams-to-grams type of problem (2 words)

Chapter 9 Answers to Self-Test

1. **a** 2. **b** 3. **a** 4. **b** 5. **c** 6. **b** 7. **c** 8. **b** 9. **a** 10. **c** 11. **c** 12. **c** 13. **b** 14. **d** 15. **b**
16. **a** 17. **d** 18. **b** 19. **c** 20. **a** 21. **b** 22. **a** 23. **c**

Stoichiometry Problems

24. $BaCl_2(s) + 2\ AgNO_3(aq) \rightarrow 2\ AgCl(s) + Ba(NO_3)_2(aq)$

$$0.333\ \cancel{g\ BaCl_2} \times \frac{1\ \cancel{mol\ BaCl_2}}{208.23\ \cancel{g\ BaCl_2}} \times \frac{2\ \cancel{mol\ AgCl}}{1\ \cancel{mol\ BaCl_2}} \times \frac{143.32\ g\ AgCl}{1\ \cancel{mol\ AgCl}} = g\ AgCl$$

$$= 0.458\ g\ AgCl$$

25. $2\ AgClO_3(s) \xrightarrow{\Delta} 2\ AgCl(s) + 3\ O_2(g)$

$$1.555\ \cancel{g\ AgClO_3} \times \frac{1\ \cancel{mol\ AgClO_3}}{191.32\ \cancel{g\ AgClO_3}} \times \frac{3\ \cancel{mol\ O_2}}{2\ \cancel{mol\ AgClO_3}} \times \frac{22.4\ L\ O_2}{1\ \cancel{mol\ O_2}} = L\ O_2$$

$$= 0.273\ L\ O_2$$

Chapter 9 Answers to Crossword Puzzle

Across
1. volume–volume
3. mass–volume
4. actual
9. iron
15. conservation of mass
16. ammonia
17. mole ratio
18. STP

Down
2. molecule
5. combining volumes
6. stoichiometry
7. limiting reactant
8. theoretical
10. percent yield
11. Avogadro
12. molar mass
13. molar volume
14. mass–mass

CHAPTER

Gases 10

Section 10.1 *Properties of Gases*

1. Which of the following is an observed property of gases?
   (a) gases have a variable shape and volume
   (b) gases expand infinitely
   (c) gases have low density
   (d) gases mix completely
   (e) all of the above

Section 10.2 *Atmospheric Pressure*

2. Which of the following expresses standard atmospheric pressure?
   (a) 29.9 in. Hg
   (b) 76 cm Hg
   (c) 760 mm Hg
   (d) 14.7 psi
   (e) all of the above

3. If the pressure of nitrogen gas is 15 mm Hg, what is the pressure in atm?
   (a) 0.020 atm
   (b) 0.20 atm
   (c) 15 atm
   (d) 1100 atm
   (e) 11,000 atm

4. If the atmospheric pressure is 30.4 in. Hg, what is the pressure in psi?
   (a) 14.9 psi
   (b) 30.4 psi
   (c) 149 psi
   (d) 3990 psi
   (e) 39,900 psi

Section 10.3 *Variables Affecting Gas Pressure*

5. The pressure exerted by a gas is affected by three factors. Which of the following *increases* the pressure?
   (a) increasing the volume
   (b) decreasing the temperature
   (c) increasing the number of gas molecules
   (d) all of the above
   (e) none of the above

Section 10.4 *Boyle's Law*

6. A sample of ammonia gas at 1.20 atm compresses from 500 mL to 250 mL. If the temperature remains constant, what is the new pressure of the gas?
   (a) 0.600 atm
   (b) 1.00 atm
   (c) 1.20 atm
   (d) 2.40 atm
   (e) 4.80 atm

7. Which of the following graphs represents the plot of pressure versus the volume at constant temperature?

(a)

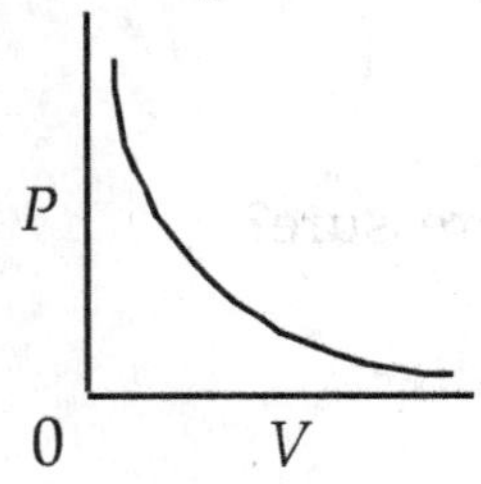

(b)

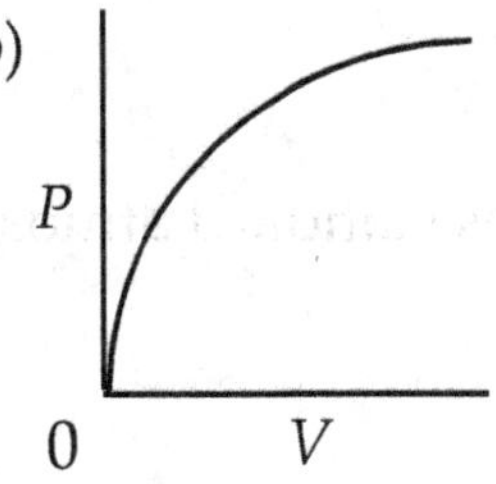

(c)

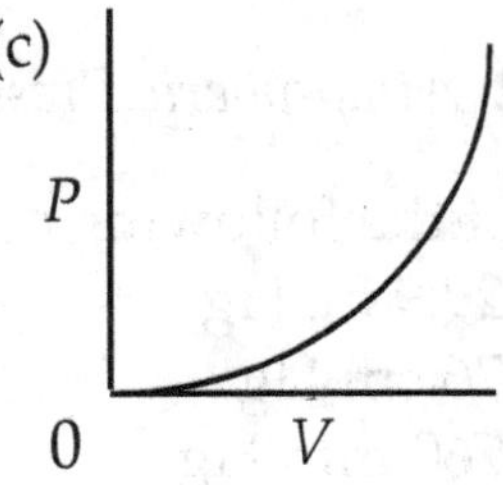

(d)

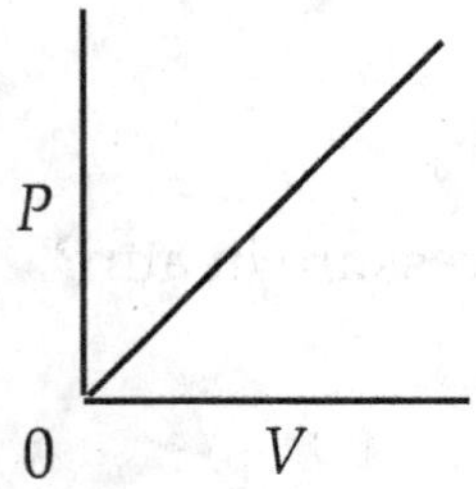

(e)

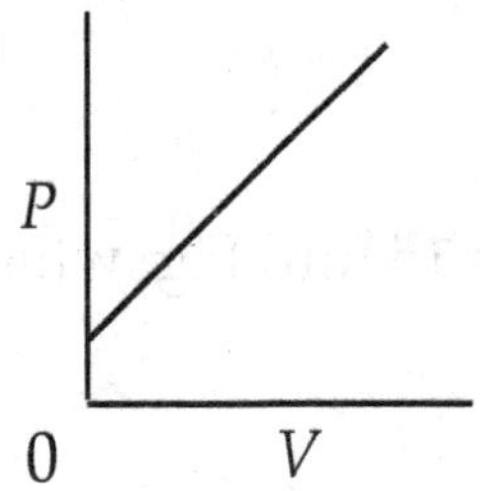

Section 10.5 *Charles's Law*

8. A sample of helium gas at 45 °C expands from 125 mL to 250 mL. If the pressure remains constant, what is the new Celsius temperature of the gas?
   (a) 23 °C
   (b) 90 °C
   (c) 159 °C
   (d) 363 °C
   (e) 636 °C

Which of the following graphs represents the plot of volume versus the Kelvin temperature at constant pressure?

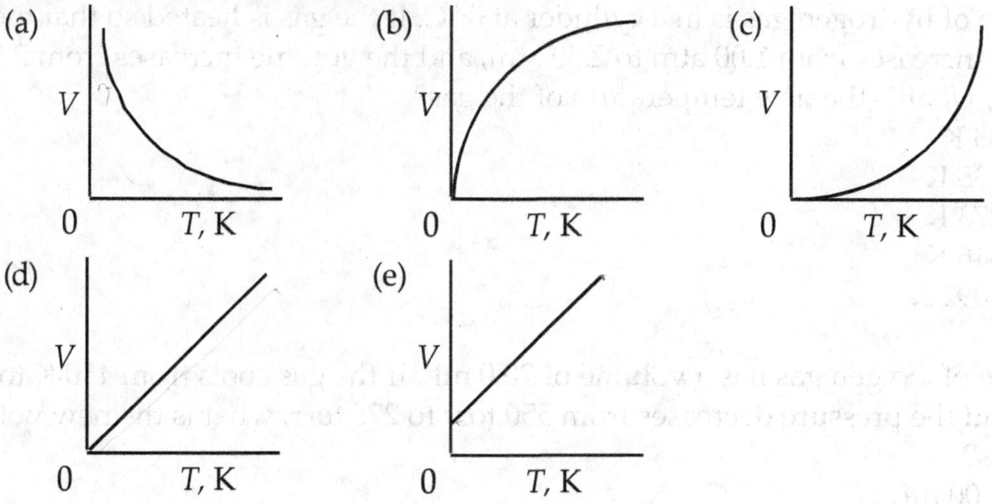

Section 10.6 *Gay-Lussac's Law*

10. A sample of chlorine gas at 785 mm Hg is heated from 50 °C to 100 °C. If the volume remains constant, what is the new pressure of the gas?
   (a) 153 mm Hg
   (b) 393 mm Hg
   (c) 680 mm Hg
   (d) 907 mm Hg
   (e) 1570 mm Hg

11. Which of the following graphs represents the plot of pressure versus the Kelvin temperature at constant volume?

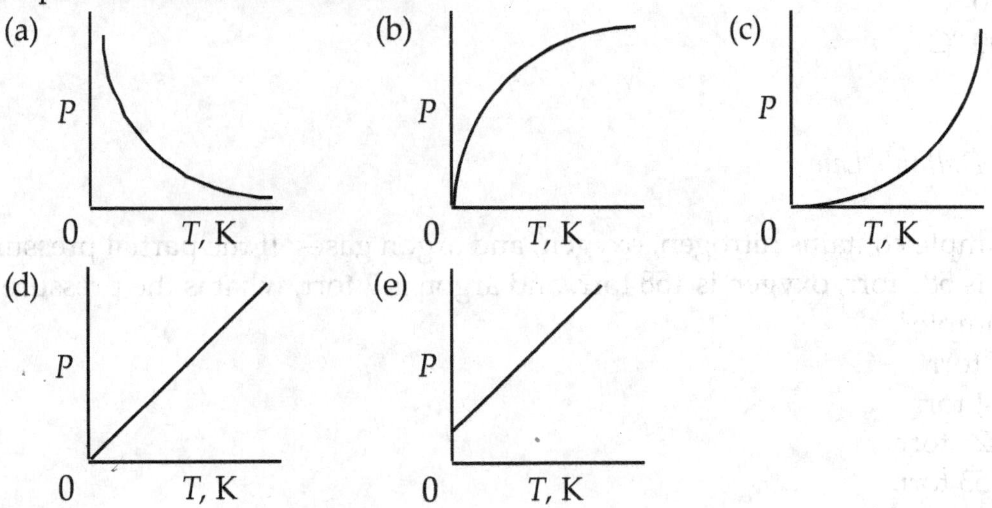

Section 10.7 *Combined Gas Law*

12. A sample of hydrogen gas is in a cylinder at 0 °C. If the gas is heated so that the pressure increases from 1.00 atm to 2.00 atm, and the volume increases from 2.50 L to 5.00 L, what is the new temperature of the gas?
    (a) 68 K
    (b) 135 K
    (c) 273 K
    (d) 546 K
    (e) 1092 K

13. A sample of oxygen gas has a volume of 20.0 mL. If the gas cools from 450 K to 225 K, and the pressure decreases from 550 torr to 275 torr, what is the new volume of the gas?
    (a) 5.00 mL
    (b) 10.0 mL
    (c) 20.0 mL
    (d) 40.0 mL
    (e) 80.0 mL

Section 10.8 *The Vapor Pressure Concept*

14. The vapor pressure of ethanol is 360 mm Hg at 60 °C; 600 mm Hg at 72 °C; 760 mm Hg at 78 °C; 810 mm Hg at 80 °C; and 1190 mm Hg at 90 °C. What is the normal boiling point of ethanol?
    (a) 60 °C
    (b) 72 °C
    (c) 78 °C
    (d) 80 °C
    (e) 90 °C

Section 10.9 *Dalton's Law*

15. An air sample contains nitrogen, oxygen, and argon gases. If the partial pressure of nitrogen is 588 torr, oxygen is 158 torr, and argon is 7 torr, what is the pressure of the air sample?
    (a) 7 torr
    (b) 14 torr
    (c) 423 torr
    (d) 753 torr
    (e) 768 torr

16. Oxygen is collected over water at 50.0 °C and 770.5 mm Hg. What is the partial pressure of the oxygen gas? (Vapor pressure of water at 50 °C is 92.5 mm Hg.)
   (a) 678.0 mm Hg
   (b) 720.5 mm Hg
   (c) 820.5 mm Hg
   (d) 863.0 mm Hg
   (e) none of the above

Section 10.10 *Ideal Gas Behavior*

17. Which of the following is *not* true according to the kinetic theory of gases?
   (a) molecules occupy zero volume
   (b) molecules are attracted to one another
   (c) molecules move randomly
   (d) molecules have elastic collisions
   (e) molecules have a kinetic energy proportional to temperature

18. At the same temperature, which of the following gases has the fastest velocity?
   (a) hydrogen
   (b) helium
   (c) oxygen
   (d) fluorine
   (e) The average velocity for each gas is the same.

19. What is the temperature at which an ideal gas exerts zero pressure?
   (a) 0 °C
   (b) 273 °C
   (c) 0 K
   (d) 273 K
   (e) -273 K

Section 10.11 *Ideal Gas Law*

20. If 1.00 mol of krypton gas exerts a pressure of 1.00 atm at 100 °C, what is the volume of the gas? (R = 0.0821 atm•L/mol•K)
   (a) 2.20 mL
   (b) 8.21 L
   (c) 12.2 L
   (d) 30.6 L
   (e) 1220 L

Combined Gas Law Problems

21. Calcium metal reacts with water to produce hydrogen gas. If 45.5 mL of hydrogen gas is collected over water at 20 °C and 752 mm Hg, what is the volume at STP? (The vapor pressure of water at 20 °C is 17.5 mm Hg.)

	P	*V*	*T*
initial			
final			

22. The decomposition of a baking soda sample by heating gives 5.50 L of carbon dioxide gas at 100 °C and 2.25 atm pressure. If the volume of carbon dioxide gas is 2.77 L at 7.50 atm, what is the temperature of the gas?

	P	*V*	*T*
initial			
final			

Chapter 10 Key Terms

Across

1. molecular collisions that do not lose energy
4. __________ by displacement
5. stated the relationship of pressure and temperature
7. a volume containing no gas molecules
9. describes the behavior of an ideal gas (2 words)
15. pressure exerted by a gas in a mixture (2 words)
19. stated the relationship of gas pressure and volume
20. *PV* =
22. stated the relationship of volume and temperature
23. *T* and *V* are __________ proportional

Down

2. stated equal volumes—same number of molecules
3. gas law that relates *P*, *V*, and *T*
6. the pressure exerted by molecules in air
8. the lowest possible temperature (2 words)
10. *P* and *V* are __________ proportional
11. 0 °C and 1 atm
12. frequency of molecular collisions (2 words)
13. instrument for measuring atmospheric pressure
14. stated pressure of gases in a mixture is additive
16. equals 1 mm Hg gas pressure
17. gas that deviates from ideal behavior
18. gas–liquid equilibrium pressure
21. gas that obeys kinetic theory

Chapter 10 Answers to Self-Test

1. e 2. e 3. a 4. a 5. c 6. d 7. a 8. d 9. d 10. d 11. d 12. e 13. c 14. c 15. d 16. a 17. b 18. a 19. c 20. d

Combined Gas Law Problems

21.

	P	V	T
initial	752 – 17.5 = 734 mm Hg	45.5 mL	20 °C + 273 = 293 K
final	760 mm Hg	V_{final}	273 K

$$45.5\ \text{mL} \times \frac{734\ \cancel{\text{mm Hg}}}{760\ \cancel{\text{mm Hg}}} \times \frac{273\ \cancel{\text{K}}}{293\ \cancel{\text{K}}} = 40.9\ \text{mL}$$

22.

	P	V	T
initial	2.25 atm	5.50 L	100 °C + 273 = 373 K
final	7.50 atm	2.77 L	T_{final}

$$373\ \text{K} \times \frac{7.50\ \cancel{\text{atm}}}{2.25\ \cancel{\text{atm}}} \times \frac{2.77\ \cancel{\text{L}}}{5.50\ \cancel{\text{L}}} = 626\ \text{K}\ (353\ °\text{C})$$

Chapter 10 Answers to Crossword Puzzle

Across

1. elastic
4. volume
5. Gay-Lussac
7. vacuum
9. kinetic theory
15. partial pressure
19. Boyle
20. nRT
22. Charles
23. directly

Down

2. Avogadro
3. combined
6. atmospheric
8. absolute zero
10. inversely
11. STP
12. gas pressure
13. barometer
14. Dalton
16. torr
17. real
18. vapor
21. ideal

CHAPTER

11

Liquids and Solids

Section 11.1 *Properties of Liquids*

1. Which of the following are observed general properties of liquids?
   (a) liquids have a variable shape and fixed volume
   (b) liquids compress and expand significantly
   (c) liquids are much less dense than gases
   (d) liquids that are insoluble mix homogeneously
   (e) none of the above

2. Predict the physical state of argon at –190 °C (Mp = –189 °C, Bp = –186 °C) and normal atmospheric pressure.
   (a) gas
   (b) liquid
   (c) solid
   (d) solid and liquid
   (e) none of the above

Section 11.2 *The Intermolecular Bond Concept*

3. What is the strongest intermolecular force in a liquid containing polar molecules?
   (a) covalent bonds
   (b) ionic bonds
   (c) permanent dipole forces
   (d) temporary dipole forces
   (e) none of the above

4. What is the strongest intermolecular force in a liquid containing nonpolar molecules?
   (a) covalent bonds
   (b) dipole forces
   (c) dispersion forces
   (d) hydrogen bonds
   (e) none of the above

5. What is the strongest intermolecular force in a liquid containing molecules with H-N bonds?
   (a) dipole forces
   (b) dispersion forces
   (c) hydrogen bonds
   (d) covalent bonds
   (e) none of the above

6. Given the following liquids have similar molar masses, which has the strongest molecular attraction?
   (a) CH_3-CO-O-H
   (b) CH_3-CH_2-O-CH_3
   (c) CH_3-CH_2-S-CH_3
   (d) CH_3-CH_2-CH_2-Cl
   (e) CH_3-CH_2-CH_2-CH_2-CH_3

Section 11.3 *Vapor Pressure, Boiling Point, Viscosity, and Surface Tension*

7. Consider the following liquids with similar molar masses. Which liquid has the strongest attraction between molecules based only on boiling point data?
   (a) Liquid X (Bp = 50 °C @ 760 mm Hg)
   (b) Liquid Y (Bp = 0 °C @ 760 mm Hg)
   (c) Liquid Z (Bp = –50 °C @ 760 mm Hg)
   (d) all liquids have the same intermolecular attraction
   (e) insufficient data to predict

8. Consider the following liquids with similar molar masses. Which liquid has the strongest attraction between molecules based on vapor pressure data?
   (a) Liquid X (vapor pressure = 50 mm Hg @ 20 °C)
   (b) Liquid Y (vapor pressure = 100 mm Hg @ 20 °C)
   (c) Liquid Z (vapor pressure = 150 mm Hg @ 20 °C)
   (d) all liquids have the same intermolecular attraction
   (e) insufficient data to predict

9. Consider the following liquids with similar molar masses. Which liquid has the strongest attraction between molecules based on viscosity data?
   (a) Liquid X (viscosity = 0.50 centipoise @ 20 °C)
   (b) Liquid Y (viscosity = 0.75 centipoise @ 20 °C)
   (c) Liquid Z (viscosity = 1.25 centipoise @ 20 °C)
   (d) all liquids have the same intermolecular attraction
   (e) insufficient data to predict

0. Consider the following liquids with similar molar masses. Which liquid has the strongest attraction between molecules based on surface tension data?
   (a) Liquid X (surface tension = 15 dynes/cm @ 20 °C)
   (b) Liquid Y (surface tension = 25 dynes/cm @ 20 °C)
   (c) Liquid Z (surface tension = 50 dynes/cm @ 20 °C)
   (d) all liquids have the same intermolecular attraction
   (e) insufficient data to predict

Section 11.4 *Properties of Solids*

11. Which of the following properties is *not* a general characteristic of solids?
   (a) crystalline and noncrystalline structures
   (b) do not compress or expand significantly
   (c) heterogeneous solids mix by diffusion
   (d) rigid shape and fixed volume
   (e) usually more dense than corresponding liquids

12. Predict the physical state of argon at –180 °C (Mp = –189 °C, Bp = –186 °C) and normal atmospheric pressure.
   (a) gas
   (b) liquid
   (c) solid
   (d) solid and liquid
   (e) liquid and gas

Section 11.5 *Crystalline Solids*

13. Which of the following is an example of an ionic crystalline solid?
   (a) calcite, $CaCO_3$
   (b) fluorite, CaF_2
   (c) iron pyrite, FeS_2
   (d) silver chloride, $AgCl$
   (e) all of the above

14. Which of the following is *not* an example of a molecular crystalline solid?
   (a) dry ice, CO_2
   (b) rose gold, Au/Cu
   (c) sucrose, $C_{12}H_{22}O_{11}$
   (d) sulfur, S_8
   (e) urea, $CO(NH_2)_2$

15. Which of the following is an example of a metallic crystalline solid?
   (a) diamond, C
   (b) halite, NaCl
   (c) iodine, I_2
   (d) phosphorus, P_4
   (e) solder, Pb/Sn

Section 11.6 *Changes of Physical State*

16. Calculate the heat required to convert 10.0 g of ice at –10.0 °C to water at 20 °C. The specific heat of ice is 0.50 cal/(g × °C); the heat of fusion for ice is 80.0 cal/g; and the specific heat of water is 1.00 cal/(g × °C).
    (a) 105 cal
    (b) 825 cal
    (c) 1050 cal
    (d) 1200 cal
    (e) 2800 cal

17. Calculate the heat released when 25.0 g of steam at 100.0 °C cools to water at 0 °C. The heat of condensation for steam is 540.0 cal/g; and the specific heat of water is 1.00 cal/(g × °C).
    (a) 4500 cal
    (b) 13,500 cal
    (c) 15,500 cal
    (d) 16,000 cal
    (e) 18,000 cal

Section 11.7 *Structure of Water*

18. What is the number of nonbonding electron pairs in a water molecule?
    (a) 0 (b) 1 (c) 2 (d) 3 (e) 4

19. What is the experimentally observed bond angle in a water molecule?
    (a) 90°
    (b) 104.5°
    (c) 109.5°
    (d) 120°
    (e) 180°

20. Which of the following illustrates the bond polarity in a water molecule?
    (a) (δ–) O–H (δ+)
    (b) (δ–) O–H (δ–)
    (c) (δ+) O–H (δ+)
    (d) (δ+) O–H (δ–)
    (e) (δ) O–H (δ)

ction 11.8 *Physical Properties of Water*

21. Which of the following accounts for the unusually high heats of fusion and vaporization for water?
    (a) the hydrogen bonds in water
    (b) the molar mass of water
    (c) the specific heat of water
    (d) the surface tension in water
    (e) the viscosity of water

22. Which of the following physical properties has a higher value for water (H_2O) than the corresponding value for heavy water, (D_2O)?
    (a) density
    (b) melting point
    (c) heat of vaporization
    (d) molar mass
    (e) none of the above

Section 11.9 *Chemical Properties of Water*

23. What are the products from the complete combustion of a hydrocarbon?
    (a) carbon and water
    (b) carbon monoxide and hydrogen
    (c) carbon monoxide and water
    (d) carbon dioxide and hydrogen
    (e) carbon dioxide and water

24. Complete the following chemical equation and indicate the product(s).

$$BaO(s) + H_2O(l) \rightarrow$$

    (a) $Ba(OH)_2$
    (b) $Ba(OH)_2 + H_2$
    (c) $Ba(OH)_2 + O_2$
    (d) $Ba(OH)_2 + O_3$
    (e) $Ba(OH)_2 + H_2O_2$

25. Complete the following chemical equation and indicate the product(s).

$$CO_2(g) + H_2O(l) \rightarrow$$

    (a) H_2CO_3
    (b) $HC_2H_3O_2$
    (c) CH_3OH
    (d) $CO + H_2O_2$
    (e) $CO + H_2 + O_2$

Section 11.10 *Hydrates*

26. What is the systematic name for $Ca(NO_3)_2 \cdot 4H_2O$?
    (a) calcium nitrate dihydrate
    (b) calcium nitrate tetrahydrate
    (c) calcium nitrite dihydrate
    (d) calcium nitrite tetrahydrate
    (e) none of the above

27. What is the percentage of water in borax, $Na_2B_4O_7 \cdot 10H_2O$?
    (a) 4.72%
    (b) 8.95%
    (c) 47.2%
    (d) 82.1%
    (e) 89.5%

Chemical Reactions Involving Water

28. Complete and balance each of the following chemical reactions.

$$K(s) + H_2O(l) \rightarrow$$

$$HNO_3(aq) + Zn(OH)_2(s) \rightarrow$$

$$CH_3OH(g) + O_2(g) \xrightarrow{\text{spark}}$$

$$KAl(SO_4)_2 \cdot 12H_2O(s) \xrightarrow{\Delta}$$

Empirical Formula of a Hydrate – Show all work for the following.

29. An unknown hydrate of chromium(III) acetate, $Cr(C_2H_3O_2)_3 \cdot XH_2O$, is heated to give 7.29% water. Calculate the water of crystallization (X) for the salt and state the formula and name of the hydrate.

Chapter 11 Key Terms

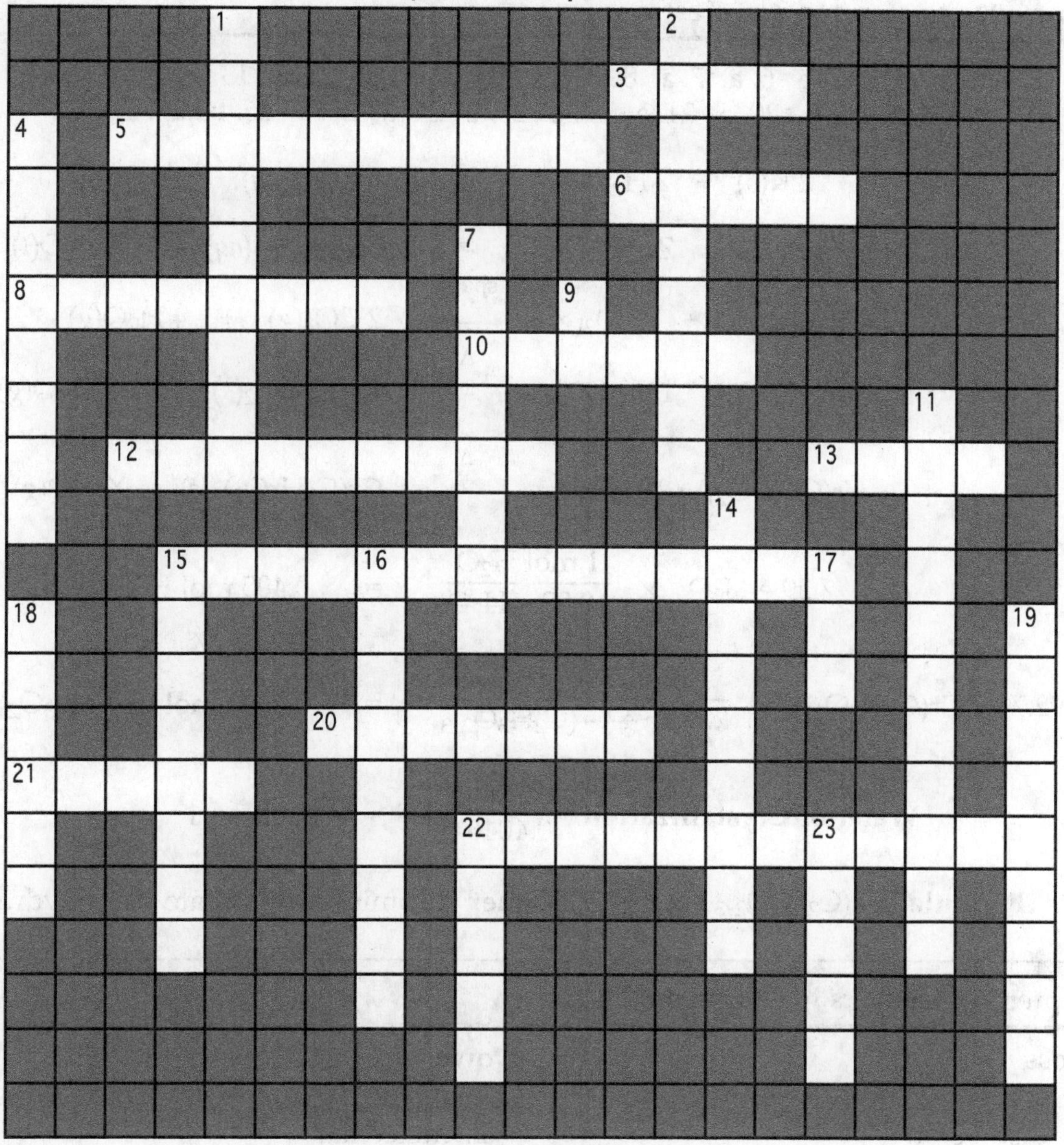

Across

3. water containing sodium ions and various anions
5. temporary attractive force between molecules
6. an oxide that gives a basic solution
8. a crystalline solid composed of metal atoms
10. permanent attractive force between molecules
12. water of ___________ (crystallization)
13. water containing a variety of cations and anions
15. the tendency of a liquid to form drops (2 words)
20. a substance containing attached water molecules
21. a crystalline solid composed of ions
22. heat required to convert a liquid to a gas

Down

1. boiled and condensed water
2. a crystalline solid composed of molecules
4. an oxide that gives an acidic solution
7. an intermolecular bond between H and O
9. temperature when vapor pressure is 1 atm
11. a solid with particles that repeat in a pattern
14. the resistance of a liquid to flow
15. heat required to raise 1 g of substance 1 °C
16. a compound that does not contain water
17. overall dipole showing partial negative charge
18. heat required to convert a solid to a liquid
19. demineralized water
22. pressure when condensation equals evaporation
23. arrangement of two atoms and the central atom

Chapter 11 Answers to Self-Test

1. a 2. c 3. c 4. c 5. c 6. a 7. a 8. a 9. c 10. c 11. c 12. a 13. e 14. b 15. e 16. c 17. d 18. c 19. b 20. a 21. a 22. e 23. e 24. a 25. a 26. b 27. c

28.

$$2\,K(s) + 2\,H_2O(l) \rightarrow 2\,KOH(aq) + H_2(g)$$

$$2\,HNO_3(aq) + Zn(OH)_2(s) \rightarrow Zn(NO_3)_2(aq) + 2\,H_2O(l)$$

$$2\,CH_3OH(g) + 3\,O_2(g) \xrightarrow{\text{spark}} 2\,CO_2(g) + 4\,H_2O(g)$$

$$KAl(SO_4)_2 \cdot 12H_2O(s) \xrightarrow{\Delta} KAl(SO_4)_2(s) + 12\,H_2O(g)$$

29.

$$Cr(C_2H_3O_2)_3 \cdot XH_2O(s) \xrightarrow{\Delta} Cr(C_2H_3O_2)_3(s) + XH_2O(g)$$

$$7.29\ \cancel{g\ H_2O} \times \frac{1\ mol\ H_2O}{18.02\ \cancel{g\ H_2O}} = 0.405\ mol\ H_2O$$

$$92.71\ \cancel{g\ Cr(C_2H_3O_2)_3} \times \frac{1\ mol\ Cr(C_2H_3O_2)_3}{229.15\ \cancel{g\ Cr(C_2H_3O_2)_3}} = 0.4046\ mol\ Cr(C_2H_3O_2)_3$$

Water of Crystallization: $\frac{0.405}{0.4046} = 1.00 \approx 1$

Formula: $Cr(C_2H_3O_2)_3 \cdot H_2O$ **Name**: chromium(III) acetate monohydrate

Chapter 11 Answers to Crossword Puzzle

Across

3. soft
5. dispersion
6. metal
8. metallic
10. dipole
12. hydration
13. hard
15. surface tension
20. hydrate
21. ionic
22. vaporization

Down

1. distilled
2. molecular
4. nonmetal
7. hydrogen
9. Bp
11. crystalline
14. viscosity
15. specific
16. anhydrous
17. net
18. fusion
19. deionized
22. vapor
23. angle

CHAPTER

12

Chemical Bonding

Section 12.1 *The Chemical Bond Concept*

1. What type of chemical bond results from the attraction between a positive metal ion and a negative nonmetal ion?
   (a) covalent bond
   (b) ionic bond
   (c) metallic bond
   (d) valence bond
   (e) none of the above

2. Predict which of the following compounds is held together by covalent bonds.
   (a) CO_2
   (b) PbO_2
   (c) SnO_2
   (d) TiO_2
   (e) all of the above

Section 12.2 *Ionic Bonds*

3. Which of the following statements is true regarding the bond between sodium and chlorine in a NaCl formula unit?
   (a) Sodium atoms lose electrons and chlorine atoms gain electrons.
   (b) Sodium and chloride ions bond by electrostatic attraction.
   (c) The ionic radius of a sodium ion is less than its atomic radius.
   (d) Breaking an ionic bond between Na^+ and Cl^- requires energy.
   (e) all of the above

4. How many valence electrons are in a bromine atom and a bromide ion?
   (a) 1 and 7, respectively
   (b) 1 and 8, respectively
   (c) 7 and 8, respectively
   (d) 7 and 10, respectively
   (e) none of the above

5. Which noble gas is isoelectronic with a potassium ion?
   (a) neon
   (b) argon
   (c) krypton
   (d) xenon
   (e) none of the above

6. Which of the following ions has the following electron configuration: $1s^2\ 2s^2\ 2p^6\ 3s^2\ 3p^6\ 4s^2\ 3d^{10}\ 4p^6$?
   (a) I^-
   (b) S^{2-}
   (c) K^+
   (d) Sr^{2+}
   (e) none of the above

7. What is the predicted electron configuration for a titanium(IV) ion, Ti^{4+}?
   (a) $1s^2\ 2s^2\ 2p^6\ 3s^2\ 3p^6$
   (b) $1s^2\ 2s^2\ 2p^6\ 3s^2\ 3p^6\ 4s^2$
   (c) $1s^2\ 2s^2\ 2p^6\ 3s^2\ 3p^6\ 4s^2\ 3d^2$
   (d) $1s^2\ 2s^2\ 2p^6\ 3s^2\ 3p^6\ 4s^2\ 3d^6$
   (e) none of the above

Section 12.3 *Covalent Bonds*

8. Which of the following statements is true regarding a covalent bond between hydrogen and fluorine atoms in a HF molecule?
   (a) Valence electrons are transferred from hydrogen to fluorine atoms.
   (b) Bonding electrons are distributed over the entire HF molecule.
   (c) The bond length is equal to the sum of the two atomic radii.
   (d) Breaking a bond between H and F atoms releases energy.
   (e) none of the above

Section 12.4 *Electron Dot Formulas of Molecules*

9. Draw the electron dot formula for silane, SiH_4. How many pairs of nonbonding electrons are in a silane molecule?
   (a) 0
   (b) 1
   (c) 2
   (d) 4
   (e) none of the above

0. Draw the electron dot formula for acetylene, C_2H_4. State the type of bonds in an acetylene molecule.
   (a) 5 single bonds
   (b) 4 single bonds and 1 triple bond
   (c) 4 single bonds and 1 double bond
   (d) 2 double bonds and 1 triple bond
   (e) none of the above

Section 12.5 *Electron Dot Formulas of Polyatomic Ions*

11. Draw the electron dot formula for the carbonate ion, CO_3^{2-}. How many pairs of nonbonding electrons are in a carbonate ion?
   (a) 3
   (b) 4
   (c) 8
   (d) 12
   (e) none of the above

12. What type of bonds are in a carbonate ion, CO_3^{2-}?
   (a) 1 single bond and 2 double bonds
   (b) 2 single bonds and 1 double bond
   (c) 2 single bonds and 1 triple bond
   (d) 3 single bonds
   (e) none of the above

Section 12.6 *Polar Covalent Bonds*

13. Which of the following is a general trend for the electronegativity of elements in the periodic table?
   (a) increases from left to right; increases from bottom to top
   (b) increases from left to right; decreases from bottom to top
   (c) decreases from left to right; increases from bottom to top
   (d) decreases from left to right; decreases from bottom to top
   (e) none of the above

14. Using the electronegativity trends in the periodic table, predict which of the following molecules contain polar covalent bonds.
   (a) hydrogen chloride, HCl
   (b) hydrogen fluoride, HF
   (c) carbon monoxide, CO
   (d) nitrogen monoxide, NO
   (e) all of the above

15. Given the electronegativity values of I (2.5) and Br (2.8), illustrate the bond polarity in an iodine monobromide molecule, IBr, using delta notation.
    (a) (δ+) I—Br (δ+)
    (b) (δ+) I— Br (δ-)
    (c) (δ-) I— Br (δ+)
    (d) (δ-) I— Br (δ-)
    (e) none of the above

16. Given the electronegativity values for I (2.5) and Br (2.8), calculate the bond polarity in an iodine monobromide molecule, IBr.
    (a) -0.3
    (b) 0.3
    (c) 2.7
    (d) 5.3
    (e) -5.3

Section 12.7 *Nonpolar Covalent Bonds*

17. If the electronegativity values for H, N, O, and P are 2.1, 3.0, 3.5, and 2.1, respectively, which of the following is a nonpolar covalent molecule?
    (a) water, H_2O
    (b) laughing gas, N_2O
    (c) phosphine, PH_3
    (d) ammonia, NH_3
    (e) all of the above

18. Which of the following elements occurs naturally as a nonpolar diatomic molecule?
    (a) H_2
    (b) N_2
    (c) O_2
    (d) F_2
    (e) all of the above

Section 12.8 *Coordinate Covalent Bonds*

19. Which of the following molecules contains a coordinate covalent bond?
    (a) $HClO$
    (b) $HBrO_2$
    (c) HIO_3
    (d) $HClO_4$
    (e) all of the above

Section 12.9 *Hydrogen Bonds*

20. Which of the following describes the attraction between two H_2O molecules?
    (a) coordinate covalent bond
    (b) hydrogen bond
    (c) nonpolar covalent bond
    (d) polar covalent bond
    (e) none of the above

21. Which of the following is true of a hydrogen bond?
    (a) The bond is between a H atom and an O, N, or F atom.
    (b) The bond energy is greater than a covalent bond.
    (c) The bond length is shorter than a covalent bond.
    (d) all of the above
    (e) none of the above

Section 12.10 *Shapes of Molecules*

22. What is the electron pair geometry for a hydrogen selenide, H_2Se, molecule?
    (a) bent
    (b) linear
    (c) tetrahedral
    (d) trigonal planar
    (e) trigonal pyramidal

23. What is the molecular shape of a hydrogen selenide, H_2Se, molecule?
    (a) bent
    (b) linear
    (c) tetrahedral
    (d) trigonal planar
    (e) trigonal pyramidal

24. What is the electron pair geometry for an arsine, AsH_3, molecule?
    (a) bent
    (b) linear
    (c) tetrahedral
    (d) trigonal planar
    (e) trigonal pyramidal

25. What is the molecular shape of an arsine, AsH_3, molecule?
    (a) bent
    (b) linear
    (c) tetrahedral
    (d) trigonal planar
    (e) trigonal pyramidal

26. What is the electron pair geometry for a silicate, SiO_3^{2-}, polyatomic ion?
   (a) bent
   (b) linear
   (c) tetrahedral
   (d) trigonal planar
   (e) trigonal pyramidal

27. What is the molecular shape of a silicate, SiO_3^{2-}, polyatomic ion?
   (a) bent
   (b) linear
   (c) tetrahedral
   (d) trigonal planar
   (e) trigonal pyramidal

Electron Dot Formulas for Molecules and Polyatomic Ions

28. Calculate the total number of valence electrons for each of the following molecules and draw the electron dot and structural formulas.

   (a) PI_3

   (b) HOCN

29. Calculate the total number of valence electrons for each of the following polyatomic ions and draw the electron dot and structural formulas.

   (a) SO_3^{2-}

   (b) BO_3^{3-}

Chapter 12 Key Terms

Across

1. a chloride ion but *not* a chlorite ion
7. the rule involving eight valence electrons
8. valence electrons that are not shared
10. a bond resulting from the attraction of ions
11. the notation that indicates bond polarity
12. a covalent bond with equal sharing
14. distance between nuclei in a covalent bond
17. the ability to attract shared electrons
20. trigonal pyramidal is a molecular ________
22. a nitrate ion but *not* a nitride ion
26. a theory that explains molecular shape
28. electrons in *s* and *p* sublevels
29. a bond containing two electron pairs

Down

2. a bond containing one electron pair
3. valence electrons that are shared
4. a covalent bond with both electrons from one atom
5. formed by two atoms bonded to a central atom
6. a two-atom molecule
9. a covalent bond with unequal sharing
13. representative particle in an ionic compound
15. a formula that shows bonds using dashes
16. a formula that shows valence electrons
18. required to break a covalent bond
19. location of electron pairs about the central atom
21. an intermolecular bond
23. a bond containing three electron pairs
24. a particle held together by covalent bonds
25. a bond resulting from sharing electrons
27. the attraction between two atoms

Chapter 12 Answers to Self-Test

1. b 2. a 3. e 4. c 5. b 6. d 7. a 8. b 9. a 10. c 11. c 12. b 13. a 14. e 15. b 16. b 17. c 18. e 19. e 20. b 21. a 22. c 23. a 24. c 25. e 26. d 27. d

Electron Dot Formulas for Molecules and Polyatomic Ions

28.

	Molecule	Valence Electrons	Electron Dot	Structural
(a)	**PI_3**	$5 + 3(7) = 26\ e^-$	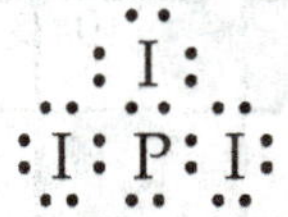	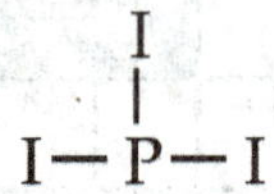
(b)	HOCN	$1 + 6 + 4 + 5 = 16\ e^-$	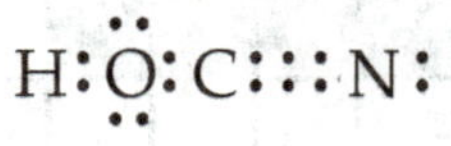	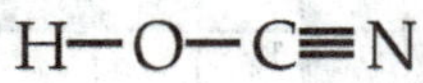

29.

	Ion	Valence Electrons	Electron Dot	Structural
(a)	**SO_3^{2-}**	$4(6) + 2 = 26\ e^-$	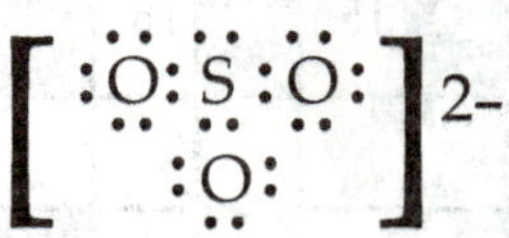	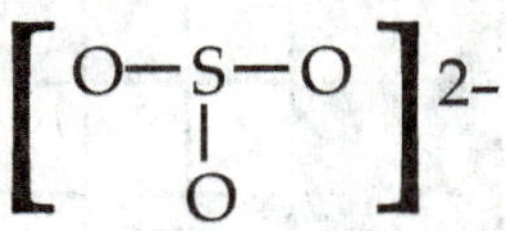
(b)	**BO_3^{3-}**	$3 + 3(6) + 3 = 24\ e^-$	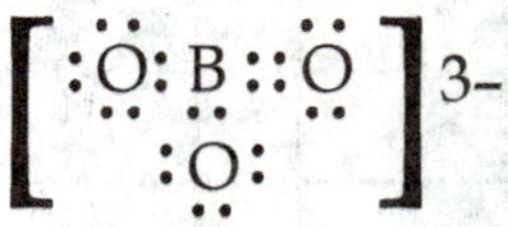	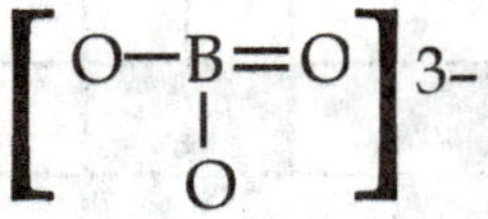

Chapter 12 Answers to Crossword Puzzle

Across

1. monoatomic
7. octet
8. nonbonding
10. ionic
11. delta
12. nonpolar
14. length
17. electronegativity
20. shape
22. polyatomic
26. VSEPR
28. valence
29. double

Down

2. single
3. bonding
4. coordinate
5. angle
6. diatomic
9. polar
13. unit
15. structural
16. dot
18. energy
19. geometry
21. hydrogen
23. triple
24. molecule
25. covalent
27. bond

CHAPTER

Solutions 13

Section 13.1 *Gases in Solution*

1. Carbonated drinks are injected with carbon dioxide gas. Under what conditions does carbon dioxide gas have the greatest solubility?
   (a) low temperature, low pressure
   (b) low temperature, high pressure
   (c) low temperature, pressure is not a factor
   (d) high pressure, temperature is not a factor
   (e) high temperature, high pressure

2. If the solubility of nitrogen gas in blood is 1.90 mL per 100 mL at one atmosphere, what is the solubility of nitrogen gas in a scuba diver's blood at a depth of 155 feet where the pressure is 5.50 atmospheres?
   (a) 0.190 mL/100 mL
   (b) 0.345 mL/100 mL
   (c) 1.90 mL/100 mL
   (d) 3.80 mL/100 mL
   (e) 10.5 mL/100 mL

Section 13.2 *Liquids in Solution*

3. Which of the following illustrates the *like dissolves like* rule for two liquids?
   (a) a polar solvent is insoluble in a polar solvent
   (b) a polar solvent is soluble in a nonpolar solvent
   (c) a nonpolar solvent is soluble in a nonpolar solvent
   (d) all of the above
   (e) none of the above

4. Apply the *like dissolves like* rule to predict which of the following liquids is miscible with water.
   (a) chloroform, $CHCl_3$
   (b) glycerin, $C_3H_5(OH)_3$
   (c) toluene, $C_6H_5CH_3$
   (d) all of the above
   (e) none of the above

Section 13.3 *Solids in Solution*

5. Which of the following illustrates the *like dissolves like* rule for a solid solute in a liquid solvent?
   (a) an ionic compound is insoluble in a polar solvent
   (b) a polar compound is insoluble in a nonpolar solvent
   (c) a nonpolar compound is insoluble in a nonpolar solvent
   (d) all of the above
   (e) none of the above

6. Apply the *like dissolves like* rule to predict which of the following vitamins is insoluble in water.
   (a) thiamine, vitamin B_1, $C_{12}H_{18}Cl_2N_4OS$
   (b) riboflavin, vitamin B_2, $C_{17}H_{20}N_4O_6$
   (c) ascorbic acid, vitamin C, $C_6H_8O_6$
   (d) calciferol, vitamin D, $C_{27}H_{44}O$
   (e) all of the above

7. Apply the *like dissolves like* rule to predict which of the following is insoluble in benzene, C_6H_6.
   (a) benzopyrene (a compound in charcoal), $C_{20}H_{12}$
   (b) DDT (an insecticide), $C_{14}H_9Cl_5$
   (c) glycine (an amino acid), $CH_2(NH_2)COOH$
   (d) naphthalene (moth repellent), $C_{10}H_8$
   (e) paradichlorobenzene (fumigant), $C_6H_4Cl_2$

Section 13.4 *The Dissolving Process*

8. When fructose, $C_6H_{12}O_6$, dissolves in water, which of the following is formed in solution?
   (a) acetic acid, $HC_2H_3O_2$
   (b) carbonic acid, H_2CO_3
   (c) hydrated clusters of CO_2 molecules
   (d) hydrated clusters of $C_6H_{12}O_6$ molecules
   (e) hydrated clusters of hydrogen and oxygen molecules

9. When solid potassium fluoride dissolves in water, which of the following is one of the aqueous ions formed?

   (a) K^+ --- H–O
H
   (b) K^+ --- O–H
H
   (c) F^- --- O–H
H

   (d) all of the above (e) none of the above

Section 13.5 *Rate of Dissolving*

10. Which of the following increases the rate of dissolving for a solid solute in a solvent?
    (a) grinding the solute
    (b) heating the solution
    (c) stirring the solution
    (d) all of the above
    (e) none of the above

Section 13.6 *Solubility and Temperature*

11. What is the solubility of sucrose sugar, $C_{12}H_{22}O_{11}$, at 20 °C? *(Refer to textbook Figure 13.5.)*
    (a) 55 g/100 g water
    (b) 75 g/100 g water
    (c) 90 g/100 g water
    (d) 100 g/100 g water
    (e) 130 g/100 g water

12. What is the minimum temperature required to dissolve 140 g of sucrose sugar, $C_{12}H_{22}O_{11}$, in 100 g of water? *(Refer to textbook Figure 13.5.)*
    (a) 20 °C
    (b) 50 °C
    (c) 55 °C
    (d) 60 °C
    (e) 100 °C

Section 13.7 *Unsaturated, Saturated, and Supersaturated Solutions*

13. How concentrated is a solution containing 100 g of $NaC_2H_3O_2$ in 100 g water at 75 °C? *(Refer to textbook Figure 13.6.)*
    (a) unsaturated
    (b) saturated
    (c) supersaturated
    (d) superunsaturated
    (e) none of the above

14. How concentrated is a solution containing 100 g of $NaC_2H_3O_2$ in 100 g water at 25 °C? *(Refer to textbook Figure 13.6.)*
    (a) unsaturated
    (b) saturated
    (c) supersaturated
    (d) superunsaturated
    (e) none of the above

Section 13.8 *Mass/Mass Percent Concentration*

15. Which of the following is *not* a unit factor related to a 5.00% aqueous solution of sodium hydroxide, NaOH?

(a) $\frac{5.00\text{ g NaOH}}{95.00\text{ g solution}}$ (b) $\frac{95.00\text{ g water}}{5.00\text{ g NaOH}}$

(c) $\frac{5.00\text{ g NaOH}}{95.00\text{ g water}}$ (d) $\frac{100\text{ g solution}}{5.00\text{ g NaOH}}$

(e) $\frac{95.00\text{ g water}}{100\text{ g solution}}$

16. What is the mass of a 10.0% blood plasma solution that contains 2.50 g of dissolved solute?
    (a) 0.250 g
    (b) 0.278 g
    (c) 22.5 g
    (d) 25.0 g
    (e) 250 g

17. What is the mass of glucose, $C_6H_{12}O_6$, solute in 10.0 g of a 5.00% sugar solution?
    (a) 0.180 g
    (b) 0.500 g
    (c) 0.900 g
    (d) 9.50 g
    (e) 10.0 g

18. What is the mass of water needed to prepare 5.00 kg of a 40.0% antifreeze solution?
    (a) 2.00 kg
    (b) 3.00 kg
    (c) 3.33 kg
    (d) 12.5 kg
    (e) 200 kg

19. An hour after a meal, a patient's blood glucose level is 120 mg/dL. Express this glucose level as mass/mass percent concentration. (Assume the density of blood is 1.0 g/mL.)
    (a) 12%
    (b) 1.2%
    (c) 0.12%
    (d) 0.012%
    (e) 0.0012%

Section 13.9 *Molar Concentration*

20. Which of the following is *not* a unit factor related to a 0.500 M NaOH solution?

(a) $\frac{0.500 \text{ mol NaOH}}{1 \text{ L solution}}$ (b) $\frac{1 \text{ L solution}}{0.500 \text{ mol NaOH}}$

(c) $\frac{0.500 \text{ mol NaOH}}{1000 \text{ mL solution}}$ (d) $\frac{1000 \text{ mL solution}}{0.500 \text{ mol NaOH}}$

(e) $\frac{0.500 \text{ mol NaOH}}{1000 \text{ g solution}}$

21. What is the mass of barium hydroxide (171.35 g/mol) dissolved in 250.0 mL of 0.200 M $Ba(OH)_2$ solution?
    (a) 8.57 g
    (b) 17.1 g
    (c) 85.7 g
    (d) 171 g
    (e) 857 g

22. What volume of 6.00 M sulfuric acid contains 9.80 g of H_2SO_4 (98.03 g/mol)?
    (a) 0.600 mL
    (b) 16.7 mL
    (c) 60.0 mL
    (d) 167 mL
    (e) 1670 mL

Section 13.10 *Dilution of a Solution*

23. Dilute sulfuric acid is a 6.0 M solution. What is the molarity of a H_2SO_4 solution prepared by diluting 50.0 mL of dilute acid to a total volume of 1.00 L?
    (a) 0.10 M
    (b) 0.12 M
    (c) 0.30 M
    (d) 0.50 M
    (e) 3.33 M

Section 13.11 *Solution Stoichiometry*

24. Given that 25.0 mL of 0.100 M barium chloride solution reacts completely with 22.5 mL of silver nitrate solution, what is the molarity of the $AgNO_3$ solution?

$$BaCl_2(aq) + 2\,AgNO_3(aq) \rightarrow 2\,AgCl(s) + Ba(NO_3)_2(aq)$$

    (a) 0.0450 M
    (b) 0.0556 M
    (c) 0.111 M
    (d) 0.222 M
    (e) 0.180 M

Mass/Mass Percent and Molar Concentrations – Show all work for the following.

25. A 50.0-mL sample of potassium iodide solution has a mass of 53.810 g and contains 5.379 g of solid KI. Calculate the mass/mass percent concentration of the solution.

26. Given the data in the preceding problem, calculate the molar concentration of the KI solution.

Chapter 13 Key Terms

Across

2. a solvent composed of polar molecules
3. liquids that dissolve in one another
5. the lesser component of a solution
7. a solvent composed of nonpolar molecules
8. stated gas solubility is proportional to pressure
11. mol solute/L of solution concentration
12. a mixture of 1–100 nm dispersed particles
16. a solution with more than maximum solute
19. a colloidal solution
20. the overall direction of negative charge (2 words)
21. *like* _________ *like* rule

Down

1. the greater component of a solution
4. relates quantities according to a balanced equation
6. a solution with less than maximum solute
9. liquids that do not dissolve in one another
10. a solute dissolved in a solvent
13. a region with partial negative and positive charge
14. solvent molecules surrounding a solute particle
15. a solution with the maximum solute
16. amount of solute that will dissolve in a solvent
17. g solute/100 g of solution concentration
18. type of light scattering by colloidal particles

Chapter 13 Answers to Self-Test

1. **b** 2. **e** 3. **c** 4. **b** 5. **b** 6. **d** 7. **c** 8. **d** 9. **b** 10. **d** 11. **d** 12. **c** 13. **a** 14. **c** 15. **a**
16. **d** 17. **b** 18. **b** 19. **c** 20. **e** 21. **a** 22. **b** 23. **c** 24. **d**

Mass/Mass Percent and Molar Concentrations

25. $$\frac{5.379 \text{ g KI}}{53.810 \text{ g soln}} \times 100\% = \text{m/m \% KI}$$

$$= 9.996\% \text{ KI}$$

26. $$\frac{5.379 \cancel{\text{g KI}}}{50.0 \cancel{\text{mL soln}}} \times \frac{1 \text{ mol KI}}{166.00 \cancel{\text{g KI}}} \times \frac{1000 \cancel{\text{mL soln}}}{1 \text{ L soln}} = \text{mol KI/L soln}$$

$$= 0.648\ M \text{ KI}$$

Chapter 13 Answers to Crossword Puzzle

Across
2. polar
3. miscible
5. solute
7. nonpolar
8. Henry
11. molarity
12. colloid
16. supersaturated
19. dispersion
20. net dipole
21. dissolves

Down
1. solvent
4. stoichiometry
6. unsaturated
9. immiscible
10. solution
13. dipole
14. cage
15. saturated
16. solubility
17. percent
18. Tyndall

CHAPTER

14

Acids and Bases

Section 14.1 *Properties of Acids and Bases*

1. Which of the following is a general property of an acidic solution?
   (a) tastes sour
   (b) turns litmus paper red
   (c) pH less than 7
   (d) reacts with a basic solution to give a salt and water
   (e) all of the above

2. If a solution has a pH of 11, which of the following describes the solution?
   (a) strongly acidic
   (b) weakly acidic
   (c) neutral
   (d) weakly basic
   (e) strongly basic

Section 14.2 *Arrhenius Acids and Bases*

3. Given the ionization, which of the following is a strong Arrhenius acid?
   (a) $HClO_2(aq)$ (~ 1% ionized)
   (b) $H_2SO_3(aq)$ (~ 1% ionized)
   (c) $H_3PO_4(aq)$ (~ 1% ionized)
   (d) all of the above
   (e) none of the above

4. What acid and base are neutralized to give the salt potassium chloride?
   (a) $HCl(aq)$ and $KOH(aq)$
   (b) $HCl(aq)$ and $KCl(aq)$
   (c) $HOH(aq)$ and $KCl(aq)$
   (d) $HOH(aq)$ and $KOH(aq)$
   (e) none of the above

Section 14.3 *Brønsted–Lowry Acids and Bases*

5. In the following reaction, which reactant is acting as a Brønsted–Lowry acid?

$$Na_2HPO_4(aq) + H_2CO_3(aq) \rightarrow NaHCO_3(aq) + NaH_2PO_4(aq)$$

   (a) Na_2HPO_4
   (b) H_2CO_3
   (c) $NaHCO_3$
   (d) NaH_2PO_4
   (e) H_2O

6. In the following reaction, which reactant is acting as a Brønsted–Lowry base?

$$NaHCO_3(aq) + NaH_2PO_4(aq) \rightarrow Na_2HPO_4(aq) + H_2CO_3(aq)$$

   (a) $NaHCO_3$
   (b) NaH_2PO_4
   (c) Na_2HPO_4
   (d) H_2CO_3
   (e) H_2O

Section 14.4 *Acid–Base Indicators*

7. Which of the following acid–base indicators is red in an acidic solution and yellow in a basic solution?
   (a) methyl red
   (b) bromthymol blue
   (c) phenolphthalein
   (d) all of the above
   (e) none of the above

8. Which of the following acid–base indicators is yellow in an acidic solution and blue in a basic solution?
   (a) methyl red
   (b) bromthymol blue
   (c) phenolphthalein
   (d) all of the above
   (e) none of the above

9. Which of the following acid–base indicators is colorless in an acidic solution and pink in a basic solution?
   (a) methyl red
   (b) bromthymol blue
   (c) phenolphthalein
   (d) all of the above
   (e) none of the above

Section 14.5 *Acid–Base Titrations*

10. If 10.0 mL of 0.500 M $HC_2H_3O_2$ is titrated with 0.250 M NaOH, what volume of sodium hydroxide solution is required to neutralize the acid?

$$HC_2H_3O_2(aq) + NaOH(aq) \rightarrow NaC_2H_3O_2(aq) + H_2O(l)$$

(a) 5.00 mL
(b) 10.0 mL
(c) 20.0 mL
(d) 40.0 mL
(e) 80.0 mL

11. If a 0.500 M $HC_2H_3O_2$ solution has a density of 1.00 g/mL, what is the mass percent concentration of the acetic acid solution?
(a) 2.00%
(b) 3.00%
(c) 5.00%
(d) 10.0%
(e) 50.0%

Section 14.6 *Acid–Base Standardization*

12. What is the molarity of a potassium hydroxide solution if 25.00 mL of KOH is required to neutralize 2.042 g of $KHC_8H_4O_4$ (204.23 g/mol)?

$$KHC_8H_4O_4(aq) + KOH(aq) \rightarrow K_2C_8H_4O_4(aq) + H_2O(l)$$

(a) 0.200 M
(b) 0.250 M
(c) 0.400 M
(d) 0.500 M
(e) 0.800 M

13. Vitamin C has the chemical name ascorbic acid and can be abbreviated HAsc. If 32.00 mL of 0.400 M NaOH neutralizes 2.240 g of ascorbic acid, what is the molar mass of vitamin C?

$$HAsc(aq) + NaOH(aq) \rightarrow NaAsc(aq) + H_2O(l)$$

(a) 28.0 g/mol
(b) 87.4 g/mol
(c) 175 g/mol
(d) 180 g/mol
(e) 350 g/mol

Section 14.7 *Ionization of Water*

14. Which of the following is the ionization constant expression for water?
    (a) $K_w = [H_2O]\ [H_2O]$
    (b) $K_w = [H_2O]\ [H_3O^+]\ /\ [OH^-]$
    (c) $K_w = [H_2O]\ [OH^-]\ /\ [H_3O^+]$
    (d) $K_w = [H_3O^+]\ [OH^-]$
    (e) $K_w = [H_3O^+]\ [OH^-]\ /\ [H_2O]$

15. Given an aqueous solution in which the $[H^+] = 5.0 \times 10^{-5}\ M$, what is the molar hydroxide ion concentration?
    (a) $[OH^-] = 2.0 \times 10^{-8}\ M$
    (b) $[OH^-] = 2.0 \times 10^{-9}\ M$
    (c) $[OH^-] = 2.0 \times 10^{-10}\ M$
    (d) $[OH^-] = 5.0 \times 10^{-9}\ M$
    (e) $[OH^-] = 5.0 \times 10^{-10}\ M$

16. Which of the following is true if the hydrogen ion concentration of an aqueous solution increases?
    (a) the pH increases
    (b) the K_w increases
    (c) the K_w decreases
    (d) the hydroxide ion concentration increases
    (e) the hydroxide ion concentration decreases

Section 14.8 *The pH Concept*

17. What is the pH of a solution with a 0.001 molar hydrogen ion concentration?
    (a) 1
    (b) 2
    (c) 3
    (d) 4
    (e) 10

18. What is the molar hydrogen ion concentration of stomach acid that registers a pH of 2 on a strip of pH paper?
    (a) 0.02 *M*
    (b) 0.01 *M*
    (c) 0.2 *M*
    (d) 0.1 *M*
    (e) 2 *M*

Section 14.9 *Advanced pH Calculations*

19. What is the pH of a sample solution if the $[H^+] = 0.000\ 065\ M$?
    (a) 0.81
    (b) 4.19
    (c) 4.81
    (d) 5.19
    (e) 7.81

20. What is the $[H^+]$ of a sample solution if the pH is 10.30?
    (a) 1.01 *M*
    (b) 0.000 000 000 20 *M*
    (c) 0.000 000 000 50 *M*
    (d) 0.000 000 000 020 *M*
    (e) 0.000 000 000 050 *M*

Section 14.10 *Strong and Weak Electrolytes*

21. If a light bulb in a conductivity apparatus glows brightly when testing a solution, which of the following is true?
    (a) The solution is highly ionized.
    (b) The solution is slightly ionized.
    (c) The solution is nonionized.
    (d) The solution is highly reactive.
    (e) The solution is slightly reactive.

22. Which of the following aqueous solutions are strong electrolytes?
    (a) strong acids
    (b) strong bases
    (c) soluble salts
    (d) all of the above
    (e) none of the above

23. Sodium chloride is a soluble salt and silver chloride is a slightly soluble salt. Which of the following correctly portrays aqueous solutions of the two salts?
    (a) $NaCl(aq)$ and $AgCl(s)$
    (b) $Na^+(aq) + Cl^-(aq)$ and $AgCl(s)$
    (c) $NaCl(aq)$ and $Ag^+(aq) + Cl^-(aq)$
    (d) $Na^+(aq) + Cl^-(aq)$ and $Ag^+(aq) + Cl^-(aq)$
    (e) none of the above

Section 14.11 *Net Ionic Equations*

24. Write a balanced net ionic equation given the following total ionic equation:

$$H^+(aq) + NO_3^-(aq) + NH_4OH(aq) \rightarrow NH_4^+(aq) + NO_3^-(aq) + H_2O(l)$$

(a) $HNO_3(aq) + NH_4OH(aq) \rightarrow NH_4NO_3(aq) + H_2O(l)$
(b) $H^+(aq) + OH^-(aq) \rightarrow H_2O(l)$
(c) $NO_3^-(aq) + NH_4^+(aq) \rightarrow NH_4NO_3(aq)$
(d) $H^+(aq) + NH_4OH(aq) \rightarrow NH_4^+(aq) + H_2O(l)$
(e) $HNO_3(aq) + OH^-(aq) \rightarrow NO_3^-(aq) + H_2O(l)$

Total and Net Ionic Equations

25. Write the total and net ionic equations for the following reaction.

$$H_2SO_4(aq) + 2\,NaOH(aq) \rightarrow Na_2SO_4(aq) + 2\,H_2O(l)$$

Acid–Base Titration – Show all work for the following.

26. A 0.950-g sample of sodium carbonate is titrated with hydrochloric acid using bromcresol green indicator. Calculate the molar concentration of the acid if 38.10 mL of HCl are required to completely neutralize the sodium carbonate.

$$HCl(aq) + Na_2CO_3(s) \rightarrow NaCl(aq) + H_2O(l) + CO_2(g)$$

Chapter 14 Key Terms

Across

2. the ionization equilibrium constant for water
4. dissolving process of a polar compound in water
6. a Brønsted–Lowry base is a proton __________
9. ions that do not appear in the net ionic equation
11. a substance that can accept or donate a proton
13. an ionic equation showing all ionized substances
14. dissolving process of an ionic compound in water
18. an ionic equation with no spectator ions
19. a solution that resists changes in pH
20. an Arrhenius acid releases __________ ions
21. the hydrogen ion attached to a water molecule

Down

1. a Brønsted–Lowry acid is a proton __________
3. electrolyte that is a poor conductor of electricity
4. substance that changes color with pH
5. the product of H^+ and OH^- concentrations
7. the molar H^+ expressed on an exponential scale
8. product from a neutralization reaction
10. solution whose concentration is known precisely
12. an Arrhenius base releases __________ ions
15. an electrolyte that is a good conductor
16. precise delivery of a solution using a buret
17. an indicator color change during a titration

Chapter 14 Answers to Self-Test

1. e 2. d 3. e 4. a 5. b 6. a 7. a 8. b 9. c 10. c 11. b 12. c 13. c 14. d 15. c 16. e 17. c 18. b 19. b 20. e 21. a 22. d 23. b 24. d

Total Ionic and Net Ionic Equations

25. $H_2SO_4(aq) + 2\,NaOH(aq) \rightarrow Na_2SO_4(aq) + 2\,H_2O(l)$

$$2\,H^+(aq) + SO_4^{2-}(aq) + 2\,Na^+(aq) + 2\,OH^-(aq) \rightarrow 2\,Na^+(aq) + SO_4^{2-}(aq) + 2\,H_2O(l)$$

$$H^+(aq) + OH^-(aq) \rightarrow H_2O(l)$$

Acid–Base Titration

26. $2\,HCl(aq) + Na_2CO_3(s) \rightarrow 2\,NaCl(aq) + H_2O(l) + CO_2(g)$

$$0.950\,\cancel{g\,Na_2CO_3} \times \frac{1\,\cancel{mol\,Na_2CO_3}}{105.99\,\cancel{g\,Na_2CO_3}} \times \frac{2\,mol\,HCl}{1\,\cancel{mol\,Na_2CO_3}} = 0.0179\,mol\,HCl$$

$$\frac{0.0179\,mol\,HCl}{38.10\,\cancel{mL\,soln}} \times \frac{1000\,\cancel{mL\,soln}}{1\,L\,soln} = mol\,HCl/L\,soln$$

$$= 0.470\,M\ HCl$$

Chapter 14 Answers to Crossword Puzzle

Across
2. *Kw*
4. ionization
6. acceptor
9. spectator
11. amphiprotic
13. total
14. dissociation
18. net
19. buffer
20. hydrogen
21. hydronium

Down
1. donor
3. weak
4. indicator
5. constant
7. pH
8. salt
10. standard
12. hydroxide
15. strong
16. titration
17. endpoint

CHAPTER

Advanced Problem Solving 15

Section 15.1 *Advanced Problem-Solving Strategies*

1. What is the first question to ask before attempting to solve a chemical problem?
   (a) What steps are necessary to proceed from the given to the unknown?
   (b) What is the unknown quantity that is asked for?
   (c) What is the relevant given value?
   (d) What unit factors can be applied?
   (e) What units can be canceled?

2. After setting up a problem solution, what should be done before using a calculator?
   (a) write a balanced chemical equation
   (b) determine the relevant given value
   (c) refer to a reference book
   (d) select the units of the unknown
   (e) estimate the approximate answer

3. Which of the following problem-solving techniques is most appropriate for an unknown quantity that is described by an equation; for example, $PV = nRT$?
   (a) algebraic analysis
   (b) algorithm
   (c) strategy plan
   (d) unit analysis
   (e) visualization

4. What are the units of the unknown quantity for the following problem? Calculate the molar mass of a gas given that 2.59 g occupies 1.00 L.
   (a) g/L
   (b) L/g
   (c) L/mol
   (d) g/mol
   (e) mol/g

Section 15.2 *Concept Maps – Mole Relationships*

5. Which of the following strategy plans illustrates calculating a mass of oxygen gas from a given STP volume?
   (a) g O_2 → mol O_2 → L O_2
   (b) g O_2 → L O_2 → mol O_2
   (c) L O_2 → mol O_2 → g O_2
   (d) L O_2 → g O_2 → mol O_2
   (e) mol O_2 → g O_2 → L O_2

6. If 1.00×10^{22} molecules of SO_3 gas are dissolved in 150.0 mL of aqueous solution, what is the molarity of the sulfuric acid solution?
   (a) 0.100 *M*
   (b) 0.111 *M*
   (c) 0.166 *M*
   (d) 2.49 *M*
   (e) 4.01 *M*

7. Given 50.0 mL of 0.100 *M* H_2SO_4, what is the mass of SO_3 dissolved in the sulfuric acid solution?
   (a) 0.00500 g
   (b) 0.0100 g
   (c) 0.400 g
   (d) 0.490 g
   (e) 0.800 g

8. Given 50.0 mL of 0.100 *M* H_2SO_4, what is the volume of SO_3 at STP dissolved in the sulfuric acid solution?
   (a) 5.00 mL
   (b) 56.0 mL
   (c) 112 mL
   (d) 224 mL
   (e) 448 mL

Section 15.3 *Concept Maps – Stoichiometry*

9. If a sample of methane gas, CH_4, reacts completely with oxygen gas to give 1.80 g of water vapor, what mass of methane reacted?

$$CH_4(g) + O_2(g) \xrightarrow{\text{spark}} CO_2(g) + H_2O(g)$$

   (a) 0.401 g
   (b) 0.802 g
   (c) 1.60 g
   (d) 3.20 g
   (e) 6.40 g

10. Given that 0.500 g of Na_2CO_3 reacts completely with 0.100 *M* HNO_3, what is the volume of carbon dioxide gas produced at STP?

$$Na_2CO_3(s) + HNO_3(aq) \rightarrow NaNO_3(aq) + H_2O(l) + CO_2(g)$$

(a) 26.4 mL
(b) 52.8 mL
(c) 106 mL
(d) 211 mL
(e) 423 mL

11. What volume of oxygen gas undergoes reaction with ammonia, NH_3, and platinum catalyst to give 20.0 L of nitric oxide, NO? The reaction conditions remain constant at 825°C and 10.0 atm.

$$NH_3(g) + O_2(g) \xrightarrow{Pt/825^\circ C} NO(g) + H_2O(g)$$

(a) 10.0 L
(b) 16.0 L
(c) 20.0 L
(d) 25.0 L
(e) 30.0 L

12. Given that 100.0 mL of 0.125 *M* $AgNO_3$ reacts completely with 30.0 mL of aqueous Na_2CO_3, what is the molarity of the sodium carbonate solution?

$$AgNO_3(aq) + Na_2CO_3(aq) \rightarrow Ag_2CO_3(s) + NaNO_3(aq)$$

(a) 0.208 *M*
(b) 0.278 *M*
(c) 0.417 *M*
(d) 0.625 *M*
(e) 0.833 *M*

Section 15.4 *Multiple-Reaction Stoichiometry*

13. Given the three-step "blast furnace" process that converts iron ore, Fe_2O_3, into molten "pig iron":

$$3\,Fe_2O_3(s) + CO(g) \rightarrow 2\,Fe_3O_4(s) + CO_2(g)$$
$$Fe_3O_4(s) + CO(g) \rightarrow 3\,FeO(s) + CO_2(g)$$
$$FeO(s) + CO(g) \rightarrow Fe(l) + CO_2(g)$$

How many moles of iron are produced from the reaction of 1.00 mol of Fe_2O_3?

(a) 0.500 mol
(b) 0.667 mol
(c) 2.00 mol
(d) 3.00 mol
(e) none of the above

14. Given the three-step Ostwald process for manufacturing nitric acid, HNO_3:

$$4\,NH_3(g) + 5\,O_2(g) \rightarrow 4\,NO(g) + 6\,H_2O(g)$$
$$2\,NO(g) + O_2(g) \rightarrow 2\,NO_2(g)$$
$$3\,NO_2(g) + H_2O(l) \rightarrow 2\,HNO_3(aq) + NO(g)$$

What mass of ammonia is required to produce 6.30 kg of HNO_3?

(a) 0.150 kg
(b) 1.13 kg
(c) 1.70 kg
(d) 2.55 kg
(e) 3.40 kg

Section 15.5 *Advanced Problem–Solving Examples*

15. How many water molecules are in a 2.50 in.^{3} cube of ice (d = 0.917 g/cm^3)?

(a) 4.67×10^{22} molecules
(b) 1.95×10^{23} molecules
(c) 1.49×10^{24} molecules
(d) 1.26×10^{24} molecules
(e) 4.07×10^{26} molecules

16. What is the molar mass of an unknown gas if 10.0 g occupies 2.00 L at 536 K and 5.00 atm pressure? (R = 0.0821 atm•L/mol•K)

(a) 44.0 g/mol
(b) 53.6 g/mol
(c) 82.1 g/mol
(d) 110 g/mol
(e) 176 g/mol

Advanced Problem Solving – Show all work for the following.

17. Baking soda, $NaHCO_3$, and lime, $CaCO_3$, can be used to neutralize "acid spills." Given the following equations, calculate the total mass of sulfuric acid that can be neutralized with 50.0 g of baking soda and 75.0 g of lime.

$$2\,NaHCO_3(s) + H_2SO_4(aq) \rightarrow Na_2SO_4(aq) + 2\,H_2O(l) + 2\,CO_2(g)$$
$$CaCO_3(aq) + H_2SO_4(aq) \rightarrow CaSO_4(aq) + H_2O(l) + CO_2(g)$$

Chapter 15 Key Terms

Across

2. equal before and after a chemical reaction
4. g solute/100 g of solution concentration
7. all compounds have definite __________
8. a formula showing the simplest ratio of atoms
9. the mass of 1 mol of substance (2 words)
13. a substance dissolved in water (2 words)
17. mass of substance in a given volume
18. amount of substance with 6.02×10^{23} particles
19. Avogadro's hypothesis

Down

1. the scientist given credit for 6.02×10^{23}
2. a formula showing the actual ratio of atoms
3. 273 K and 760 mm Hg
5. relates quantities in a balanced chemical equation
6. $PV = nRT$ (3 words)
9. mol solute/1 L of solution concentration
10. coefficients in an equation are in a mole ______
11. the volume of 1 mol of gas at STP (2 words)
12. the gas law relating P, V, and T
14. reactant that controls the amount of product
15. P and V are _______ proportional
16. P and T are _______ proportional

Chapter 15 Answers to Self-Test

1. **b** 2. **e** 3. **a** 4. **d** 5. **c** 6. **b** 7. **c** 8. **c** 9. **b** 10. **c** 11. **d** 12. **a** 13. **c** 14. **d** 15. **d** 16. **a**

Advanced Problem-Solving

17. $2\,NaHCO_3(s) + H_2SO_4(aq) \rightarrow Na_2SO_4(aq) + 2\,H_2O(l) + 2\,CO_2(g)$

$$50.0\ \cancel{g\ NaHCO_3} \times \frac{1\ \cancel{mol\ NaHCO_3}}{84.01\ \cancel{g\ NaHCO_3}} \times \frac{1\ \cancel{mol\ H_2SO_4}}{2\ \cancel{mol\ NaHCO_3}} \times \frac{98.09\ g\ H_2SO_4}{1\ \cancel{mol\ H_2SO_4}} = g\ H_2SO_4$$

$$= 29.2\ g\ H_2SO_4$$

$CaCO_3(aq) + H_2SO_4(aq) \rightarrow CaSO_4(aq) + H_2O(l) + CO_2(g)$

$$75.0\ g\ CaCO_3 \times \frac{1\ mol\ CaCO_3}{100.09\ g\ CaCO_3} \times \frac{1\ mol\ H_2SO_4}{1\ mol\ CaCO_3} \times \frac{98.09\ g\ H_2SO_4}{1\ mol\ H_2SO_4} = g\ H_2SO_4$$

$$= 73.5\ g\ H_2SO_4$$

Total mass of sulfuric acid: $29.2\ g + 73.5\ g = 102.7\ g\ H_2SO_4$

Chapter 15 Answers to Crossword Puzzle

Across
2. mass
4. percent
7. composition
8. empirical
9. molar mass
13. aqueous solution
17. density
18. mole
19. theory

Down
1. Avogadro
2. molecular
3. STP
5. stoichiometry
6. ideal gas law
9. molarity
10. ratio
11. molar volume
12. combined
14. limiting
15. inversely
16. directly

CHAPTER

16

Chemical Equilibrium

Section 16.1 *Collision Theory*

1. Which of the following factors affects the rate of a chemical reaction?
   (a) collision frequency
   (b) collision energy
   (c) collision orientation
   (d) all of the above
   (e) none of the above

2. Which of the following increases the collision frequency of molecules?
   (a) increasing the temperature; adding a catalyst
   (b) decreasing the temperature; adding a catalyst
   (c) decreasing the concentration; adding a catalyst
   (d) increasing the concentration; decreasing the temperature
   (e) increasing the concentration; increasing the temperature

3. Which of the following increases the collision energy of molecules?
   (a) increasing the temperature
   (b) increasing the concentration
   (c) increasing the collision frequency
   (d) decreasing the volume of the container
   (e) adding a catalyst

4. Which of the following factors increases the rate of a chemical reaction?
   (a) decreasing the concentration of reactants
   (b) decreasing the temperature
   (c) adding a catalyst
   (d) all of the above
   (e) none of the above

5. Which of the following increases the amount of product from a reaction?
   (a) adding a metal catalyst
   (b) adding an acid catalyst
   (c) using an UV light catalyst
   (d) all of the above
   (e) none of the above

Section 16.2 *Energy Profiles of Chemical Reactions*

6. If the E_{act} is lowered, which of the following is always true?
   (a) the reaction is exothermic
   (b) the reaction is endothermic
   (c) the reaction proceeds faster
   (d) the reaction proceeds slower
   (e) none of the above

7. If the heat of reaction is exothermic, which of the following is always true?
   (a) the reaction rate is fast
   (b) the reaction rate is slow
   (c) the energy of the reactants is less than the products
   (d) the energy of the reactants is greater than the products
   (e) none of the above

8. Which of the following lowers the ΔH for a chemical reaction?
   (a) increasing reactant concentration
   (b) increasing product concentration
   (c) adding a catalyst
   (d) all of the above
   (e) none of the above

Section 16.3 *The Chemical Equilibrium Concept*

9. Which of the following is true before a reaction reaches chemical equilibrium?
   (a) The amount of reactants is increasing.
   (b) The amount of products is increasing.
   (c) The amount of reactants and products are constant.
   (d) The amount of reactants and products are equal.
   (e) none of the above

10. Which of the following is true after a reaction reaches chemical equilibrium?
   (a) The amount of reactants is increasing.
   (b) The amount of products is increasing.
   (c) The amount of reactants and products are constant.
   (d) The amount of reactants and products are equal.
   (e) none of the above

11. Which of the following is true before a reaction reaches chemical equilibrium?
   (a) The rate of the forward reaction is increasing; the rate of the reverse reaction is decreasing.
   (b) The rate of the forward reaction is decreasing; the rate of the reverse reaction is increasing.
   (c) The rates of the forward and reverse reactions are increasing.
   (d) The rates of the forward and reverse reactions are decreasing.
   (e) The rates of the forward and reverse reactions are equal.

Section 16.4 *General Equilibrium Constant, K_{eq}*

12. What is the equilibrium constant expression for: 3 A + B $\rightleftarrows$ 2 C?
    (a) K_{eq} = [A] [B] / [C]
    (b) K_{eq} = $[A]^3 [B] / [C]^2$
    (c) K_{eq} = [C] / [A] [B]
    (d) K_{eq} = $[C]^2 / [A]^3 [B]$
    (e) none of the above

13. What is the equilibrium constant expression for:

$$2\,CO(g) \quad + \quad O_2(g) \quad \rightleftarrows \quad 2\,CO_2(g)$$

    (a) K_{eq} = $[CO_2]^2 / [CO]^2 [O_2]$
    (b) K_{eq} = $[CO_2]^2 / [CO]^2 [O_2]^2$
    (c) K_{eq} = $[CO]^2 [O_2] / [CO_2]^2$
    (d) K_{eq} = $[CO]^2 [O_2]^2 / [CO_2]^2$
    (e) none of the above

14. What is the equilibrium constant expression for:

$$2\,S(s) \quad + \quad 3\,O_2(g) \quad \rightleftarrows \quad 2\,SO_3(g)$$

    (a) K_{eq} = $[S] [O_2]^3 / [SO_3]^2$
    (b) K_{eq} = $[S] [O_2]^2 / [SO_3]^3$
    (c) K_{eq} = $[SO_3]^2 / [S]^2 [O_2]^3$
    (d) K_{eq} = $[SO_3]^3 / [S] [O_2]^2$
    (e) K_{eq} = $[SO_3]^2 / [O_2]^3$

15. Oxygen and ozone gases are in chemical equilibrium in the upper atmosphere. Calculate K_{eq} for the reaction given the equilibrium concentrations of each gas at 20 °C: $[O_2] = 9.38 \times 10^{-3}$ and $[O_3] = 3.40 \times 10^{-15}$.

$$2\,O_3(g) \quad \rightleftarrows \quad 3\,O_2(g)$$

    (a) K_{eq} = 4.11×10^{-28}
    (b) K_{eq} = 1.40×10^{-23}
    (c) K_{eq} = 3.62×10^{-13}
    (d) K_{eq} = 2.43×10^{8}
    (e) K_{eq} = 7.14×10^{22}

Section 16.5 *Equilibria Shifts for Gases*

16. Which of the changes listed below will shift the equilibrium to the right for the following reversible reaction?

$$4\,HCl(g) + O_2(g) + \textit{heat} \rightleftarrows 2\,Cl_2(g) + 2\,H_2O(g)$$

(a) increase $[Cl_2]$
(b) decrease [HCl]
(c) decrease temperature
(d) increase pressure
(e) add a catalyst

17. Which of the changes listed below has no effect on the equilibrium for the following reversible reaction?

$$C(s) + O_2(g) \rightleftarrows CO_2(g) + \textit{heat}$$

(a) increase $[CO_2]$
(b) increase $[O_2]$
(c) decrease $[O_2]$
(d) increase temperature
(e) increase pressure

Section 16.6 *Ionization Equilibrium Constant, K_i*

18. What is the equilibrium constant expression, K_i, for the weak acid H_2A?

$$H_2A(aq) \rightleftarrows H^+(aq) + HA^-(aq)$$

(a) $K_i = [H^+][HA^-] / [H_2A]$
(b) $K_i = [H^+]^2[HA^-] / [H_2A]$
(c) $K_i = [H^+]^2[A^-] / [H_2A]$
(d) $K_i = [H_2A] / [H^+][HA^-]$
(e) $K_i = [H_2A] / [H^+]^2[A^-]$

19. If the hydrogen ion concentration of 0.100 *M* chloroacetic acid, $HC_2H_2O_2Cl$, is 0.012 *M*, what is the ionization constant for the acid?

(a) $K_i = 1.2 \times 10^{-1}$
(b) $K_i = 1.2 \times 10^{-3}$
(c) $K_i = 1.4 \times 10^{-3}$
(d) $K_i = 1.4 \times 10^{-4}$
(e) $K_i = 1.4 \times 10^{-5}$

Section 16.7 *Equilibria Shifts for Weak Acids and Bases*

20. Which of the changes listed below will shift the equilibrium to the right for the following reversible reaction in aqueous solution?

$$HC_2H_3O_2(aq) \rightleftarrows H^+(aq) + C_2H_3O_2^-(aq)$$

(a) decrease pH
(b) decrease $[HC_2H_3O_2]$
(c) add solid $NaC_2H_3O_2$
(d) add solid NaOH
(e) none of the above

21. Which of the changes listed below will have no effect on the equilibrium for the following reversible reaction in aqueous solution?

$$NH_4OH(aq) \rightleftarrows NH_4^+(aq) + OH^-(aq)$$

(a) add solid NH_4Cl
(b) add solid KCl
(c) add gaseous HCl
(d) add gaseous NH_3
(e) all of the above

Section 16.8 *Solubility Product Equilibrium Constant, K_{sp}*

22. What is the equilibrium constant expression, K_{sp}, for slightly soluble silver phosphate in aqueous solution?

$$Ag_3PO_4(s) \rightleftarrows 3\ Ag^+(aq) + PO_4^{3-}(aq)$$

(a) $K_{sp} = [Ag^+]\ [PO_4^{3-}]$
(b) $K_{sp} = [Ag^+]\ [PO_4^{3-}]^3$
(c) $K_{sp} = [Ag^+]^3\ [PO_4^{3-}]$
(d) $K_{sp} = [Ag^+]^3\ [PO_4^{3-}]\ /\ [Ag_3PO_4]$
(e) $K_{sp} = [Ag^+]\ [PO_4^{3-}]^3\ /\ [Ag_3PO_4]$

23. What is the K_{sp} for lead(II) fluoride, PbF_2, if the lead ion concentration in a saturated solution is 0.0019 *M*?

(a) $K_{sp} = 1.4 \times 10^{-5}$
(b) $K_{sp} = 7.2 \times 10^{-6}$
(c) $K_{sp} = 1.4 \times 10^{-8}$
(d) $K_{sp} = 2.7 \times 10^{-8}$
(e) $K_{sp} = 6.9 \times 10^{-9}$

Section 16.9 *Equilibria Shifts for Slightly Soluble Compounds*

24. Which of the changes listed below will shift the equilibrium to the right for the following reversible reaction?

$$CaCO_3(s) \rightleftarrows Ca^{2+}(aq) + CO_3^{2-}(aq)$$

(a) increase $[Ca^{2+}]$
(b) increase $[CO_3^{2-}]$
(c) add solid $Ca(NO_3)_2$
(d) add solid $NaNO_3$
(e) decrease pH

25. Which of the changes listed below has no effect on the equilibrium for the following reversible reaction?

$$Hg_2I_2(s) \rightleftarrows Hg_2^{2+}(aq) + 2\,I^-(aq)$$

(a) increase $[Hg_2^{2+}]$
(b) increase $[I^-]$
(c) add solid Hg_2I_2
(d) add solid $Hg_2(NO_3)_2$
(e) add solid KI

Chapter 16 Key Terms

Across

1. the speed at which reactants decrease (3 words)
3. a chemical reaction that consumes heat energy
4. equilibrium with all species in same physical state
5. stress on a reversible reaction causes a ______
6. the energy required to reach the transition state
8. equilibrium with species in different physical states
10. an equilibrium constant for a precipitate (2 words)
16. a chemical reaction that liberates heat energy

Down

2. reaction proceeding to reactants *and* products
4. energy reactants – energy products (3 words)
7. the state of highest energy on the reaction profile
9. a dynamic state for a reversible reaction
11. an equilibrium constant for a weak acid or base
12. a theory to explain the rate of a chemical reaction
13. a substance that lowers the energy of activation
14. an equilibrium constant for all reversible reactions
15. a graph of energy as a reaction progresses

Chapter 16 Answers to Self-Test

1. **d** 2. **e** 3. **a** 4. **c** 5. **e** 6. **c** 7. **d** 8. **e** 9. **b** 10. **c** 11. **b** 12. **d** 13. **a** 14. **e** 15. **e**
16. **d** 17. **e** 18. **a** 19. **c** 20. **d** 21. **b** 22. **c** 23. **d** 24. **e** 25. **c**

Chapter 16 Answers to Crossword Puzzle

Across

1. rate of reaction
3. endothermic
4. homogeneous
5. shift
6. activation
8. heterogeneous
10. solubility product
16. exothermic

Down

2. reversible
4. heat of reaction
7. transition
9. equilibrium
11. ionization
12. collision
13. catalyst
14. general
15. profile

CHAPTER

17

Oxidation and Reduction

Section 17.1 *Oxidation Numbers*

1. What is the oxidation number of iron in a length of wire?
   (a) 0
   (b) +2
   (c) +3
   (d) all of the above
   (e) none of the above
2. What is the oxidation number of greenish-yellow chlorine gas in the free state?
   (a) 0
   (b) –1
   (c) +1
   (d) +3
   (e) +5
3. What is the oxidation number of mercury in the mercurous ion, Hg_2^{2+} ?
   (a) 0
   (b) +1
   (c) +2
   (d) +4
   (e) none of the above
4. What is the oxidation number of chromium in the dichromate ion, $Cr_2O_7^{2-}$?
   (a) –2
   (b) +2
   (c) +6
   (d) +7
   (e) none of the above
5. What is the oxidation number of carbon in $Pb(HCO_3)_4$?
   (a) –4
   (b) +2
   (c) +3
   (d) +4
   (e) none of the above

6. What is the oxidation number of chlorine in aqueous $HClO_3$?
   (a) -1
   (b) +1
   (c) +3
   (d) +5
   (e) none of the above

Section 17.2 *Oxidation–Reduction Reactions*

7. What substance is oxidized in the following redox reaction?

$$Cl_2(g) \quad + \quad 2\,Br^-(aq) \quad \rightarrow \quad 2\,Cl^-(aq) \quad + \quad Br_2(l)$$

   (a) Cl_2
   (b) Br^-
   (c) Cl^-
   (d) Br_2
   (e) H_2O

8. What substance is the oxidizing agent in the preceding redox reaction?
   (a) Cl_2
   (b) Br^-
   (c) Cl^-
   (d) Br_2
   (e) H_2O

Section 17.3 *Balancing Redox Equations: Oxidation Number Method*

9. After balancing the following redox reaction, what is the coefficient of Mn?

$$Fe_2O_3(l) \quad + \quad Mn(l) \quad \rightarrow \quad MnO_2(l) \quad + \quad Fe(l)$$

   (a) 1
   (b) 2
   (c) 3
   (d) 4
   (e) none of the above

10. After balancing the following redox reaction, what is the coefficient of H^+?

$$MnO_4^-(aq) + VO_2^-(aq) + H^+(aq) \quad \rightarrow \quad Mn^{2+}(aq) + VO_3^-(aq) + H_2O(l)$$

   (a) 2
   (b) 3
   (c) 5
   (d) 6
   (e) none of the above

11. Which of the following is *not* a guideline for balancing redox equations by the oxidation number method?
   (a) Diagram the electrons lost by the substance oxidized and gained by the substance reduced.
   (b) Write a half-reaction for the substance oxidized and the substance reduced.
   (c) Place a coefficient in front of the substance oxidized that corresponds to the number of electrons gained by the substance reduced.
   (d) Place a coefficient in front of the substance reduced that corresponds to the number of electrons lost by the substance oxidized.
   (e) Verify the total number of atoms and total ionic charge is equal for reactants and products.

Section 17.4 *Balancing Redox Equations: Half-Reaction Method*

12. Which of the following is *not* a guideline for balancing redox equations in acid by the half-reaction method?
   (a) Write a half-reaction for the substance oxidized and the substance reduced.
   (b) Balance the atoms in each half-reaction; balance oxygen with water and hydrogen with H^+.
   (c) Multiply each half-reaction by a whole number so that the number of electrons lost by the substance oxidized is equal to the electrons gained by the substance reduced.
   (d) Add the two half-reactions together and cancel identical species on each side of the equation.
   (e) Verify that the total number of atoms is equal to the total ionic charge.

13. Which of the following is *not* a guideline for balancing redox equations in base by the half-reaction method?
   (a) Write a half-reaction for the substance oxidized and the substance reduced.
   (b) Balance the atoms in each half-reaction; balance oxygen with water and hydrogen with OH^-.
   (c) Multiply each half-reaction by a whole number so that the number of electrons lost by the substance oxidized is equal to the electrons gained by the substance reduced.
   (d) Add the two half-reactions together and cancel identical species on each side of the equation.
   (e) Verify that the total ionic charge is equal for reactants and products.

14. After balancing the following redox reaction in acidic solution, what is the coefficient of water?

$$MnO_4^-(aq) + Co^{2+}(aq) \rightarrow Mn^{2+}(aq) + Co^{3+}(aq)$$

(a) 2
(b) 3
(c) 4
(d) 5
(e) none of the above

15. After balancing the following redox reaction in basic solution, what is the coefficient of water?

$$MnO_4^-(aq) + Co^{2+}(aq) \rightarrow MnO_2(s) + Co^{3+}(aq)$$

(a) 2
(b) 3
(c) 5
(d) 6
(e) none of the above

Section 17.5 *Predicting Spontaneous Redox Reactions*

16. Which of the substances listed below is the *strongest* oxidizing agent given the following *spontaneous* redox reaction?

$$FeCl_3(aq) + NaI(aq) \rightarrow I_2(s) + FeCl_2(aq) + NaCl(aq)$$

(a) $FeCl_3$
(b) NaI
(c) I_2
(d) $FeCl_2$
(e) NaCl

17. Which of the substances listed below is the *strongest* reducing agent given the following *nonspontaneous* redox reaction?

$$FeCl_3(aq) + NaBr(aq) \rightarrow Br_2(l) + FeCl_2(aq) + NaCl(aq)$$

(a) $FeCl_3$
(b) NaBr
(c) Br_2
(d) $FeCl_2$
(e) NaCl

18. Given that the following redox reactions go essentially to completion, which of the metals listed below has the *greatest* tendency to undergo oxidation?

$$Co(s) + Sn^{2+}(aq) \rightarrow Sn(s) + Co^{2+}(aq)$$
$$Al(s) + Zn^{2+}(aq) \rightarrow Zn(s) + Al^{3+}(aq)$$
$$Sn(s) + Ag^{+}(aq) \rightarrow Ag(s) + Sn^{2+}(aq)$$
$$Zn(s) + Co^{2+}(aq) \rightarrow Co(s) + Zn^{2+}(aq)$$

(a) Ag
(b) Co
(c) Sn
(d) Al
(e) Zn

19. Given that the above redox reactions go essentially to completion, which of the metals listed below is the *strongest* oxidizing agent?
(a) Ag
(b) Co
(c) Sn
(d) Al
(e) Zn

Section 17.6 *Voltaic Cells*

20. Which of the statements listed below is true regarding the following redox reaction occurring in a *spontaneous* electrochemical cell?

$$Cr(s) + Ni^{2+}(aq) \rightarrow Ni(s) + Cr^{3+}(aq)$$

(a) Cr is oxidized at the anode
(b) Ni^{2+} is reduced at the anode
(c) electrons flow from the Ni electrode to the Cr electrode
(d) anions in the salt bridge flow from the Cr half-cell to the Ni half-cell
(e) all of the above

21. Which of the statements listed below is true regarding the following redox reaction occurring in a *spontaneous* electrochemical cell?

$$F_2(g) + 2\,Br^{-}(aq) \rightarrow Br_2(l) + 2\,F^{-}(aq)$$

(a) Br^{-} is oxidized at the anode
(b) F_2 is reduced at the cathode
(c) electrons flow from the anode to the cathode
(d) anions in the salt bridge flow from the F_2 half-cell to the Br_2 half-cell
(e) all of the above

22. Nickel–cadmium batteries are used in rechargeable electronic calculators. Given the *spontaneous* reaction for a discharging NiCad battery, what substance is being *reduced*?

$$Cd(s) + NiO_2(s) + 2\,H_2O(l) \rightarrow Cd(OH)_2(s) + Ni(OH)_2(s)$$

(a) Cd
(b) NiO_2
(c) H_2O
(d) $Cd(OH)_2$
(e) $Ni(OH)_2$

23. Which of the following is an example of a spontaneous electrochemical cell that has a commercial application as a common voltaic cell?
(a) alkaline dry cell
(b) mercury dry cell
(c) lead storage battery
(d) NiCad battery
(e) all of the above

Section 17.7 *Electrolytic Cells*

24. Which of the statements listed below is true regarding the following redox reaction occurring in a *nonspontaneous* electrochemical cell?

$$MgCl_2(l) \xrightarrow{\text{electricity}} Mg(l) + Cl_2(g)$$

(a) magnesium chloride is produced at the anode
(b) magnesium metal is produced at the cathode
(c) reduction half-reaction: $2\,Cl^- \rightarrow Cl_2 + 2\,e^-$
(d) oxidation half-reaction: $Mg^{2+} + 2\,e^- \rightarrow Mg$
(e) none of the above

25. Which of the statements listed below is true regarding the following redox reaction in a *nonspontaneous* electrochemical cell?

$$Br_2(l) + 2\,NaCl(aq) \xrightarrow{\text{electricity}} Cl_2(g) + 2\,NaBr(aq)$$

(a) liquid bromine is reduced at the anode
(b) gaseous chlorine is oxidized at the cathode
(c) electrons flow from the anode to the cathode
(d) bromine is a stronger oxidizing agent than chlorine
(e) none of the above

Chapter 17 Key Terms

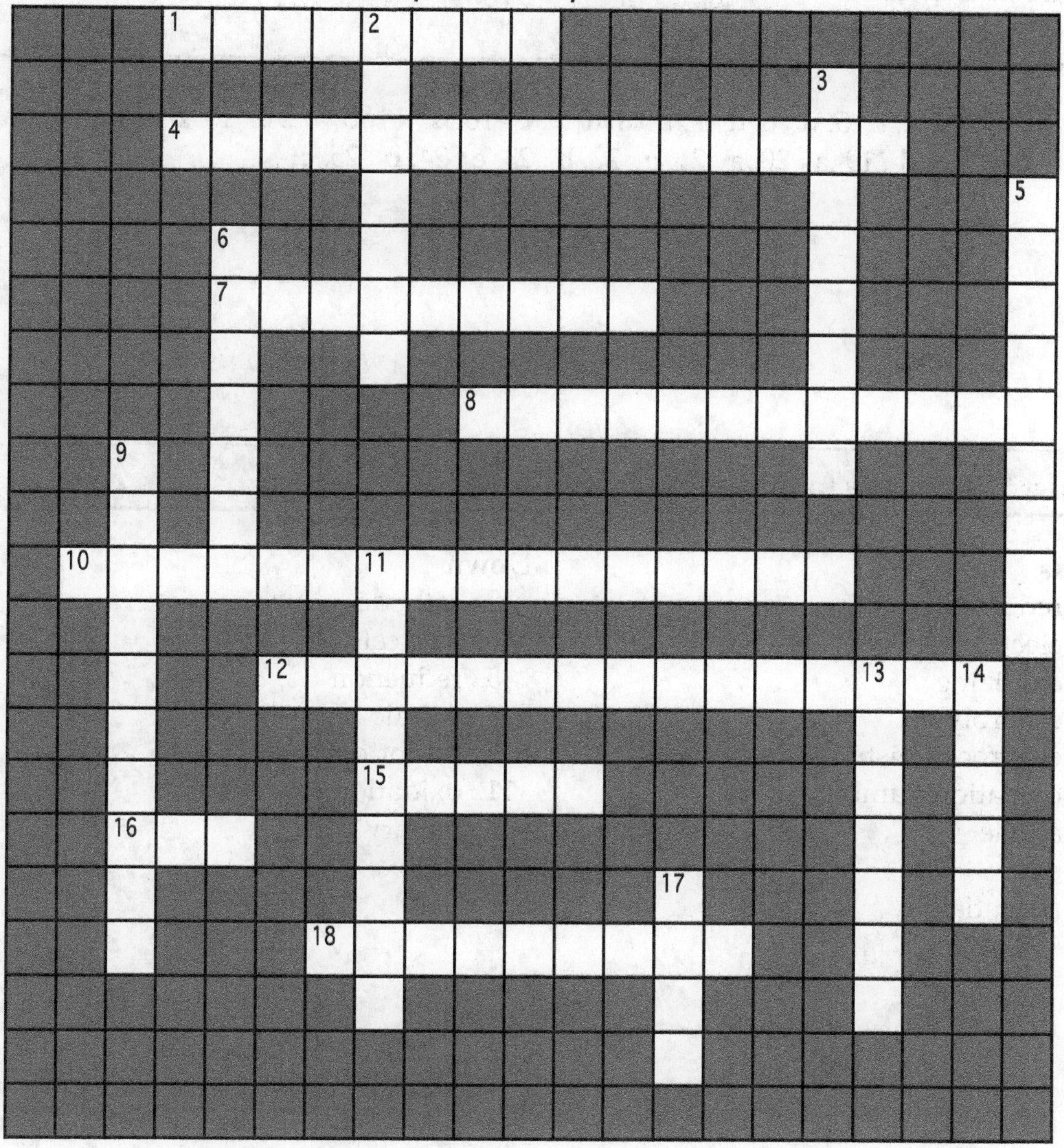

Across

1. the agent causing reduction
4. a cell having two electrodes connected by a wire
7. the agent causing oxidation
8. a nonspontaneous electrochemical cell
10. the conversion of chemical and electrical energy
12. indicates an electron rich or poor atom (2 words)
15. the electrode at which oxidation occurs
16. an electrochemical cell without aqueous solution
18. the ability of a substance to undergo reduction

Down

2. the electrode at which reduction occurs
3. a cell having a single electrode
5. a process in which a substance gains electrons
6. a spontaneous electrochemical cell
9. allows ions to travel between half-cells (2 words)
11. a process in which a substance loses electrons
13. a term for any device that produces electricity
14. a reaction in which electron transfer occurs
17. a reaction involving either oxidation or reduction

Chapter 17 Answers to Self-Test

1. a 2. a 3. b 4. c 5. d 6. d 7. b 8. a 9. c 10. d 11. b 12. e 13. b 14. c 15. a 16. a 17. d 18. d 19. a 20. a 21. e 22. b 23. e 24. b 25. c

Chapter 17 Answers to Crossword Puzzle

Across

1. reducing
4. electrochemical
7. oxidizing
8. electrolytic
10. electrochemistry
12. oxidation number
15. anode
16. dry
18. potential

Down

2. cathode
3. half-cell
5. reduction
6. voltaic
9. salt bridge
11. oxidation
13. battery
14. redox
17. half

CHAPTER

18

Nuclear Chemistry

Section 18.1 *Natural Radioactivity*

1. What are the three main types of radioactive emission from an atomic nucleus?
   (a) proton, neutron, and electron
   (b) proton, neutron, and positron
   (c) positron, neutron, and electron
   (d) positron, neutron, and electron
   (e) alpha, beta, and gamma

2. Which type of nuclear radiation is identical to a helium nucleus and is deflected toward the negative electrode as it passes between electrically charged plates?
   (a) alpha
   (b) beta
   (c) gamma
   (d) all of the above
   (e) none of the above

3. Which type of natural radiation can pass through the human body and requires thick lead protective shielding?
   (a) alpha
   (b) beta
   (c) gamma
   (d) all of the above
   (e) none of the above

Section 18.2 *Nuclear Equations*

4. What is the name of the particle having the following atomic notation: ${}^{4}_{2}He$?
   (a) alpha
   (b) beta
   (c) gamma
   (d) neutron
   (e) positron

5. What is the name of the radiation having the following atomic notation: ${}^{0}_{0}\gamma$?
   (a) alpha
   (b) beta
   (c) gamma
   (d) neutron
   (e) positron

6. What is the name of the particle having the following atomic notation: ${}^{1}_{0}n$?
   (a) alpha
   (b) electron
   (c) neutron
   (d) positron
   (e) proton

7. What is the approximate mass and relative charge of a beta particle?
   (a) 4 amu and +2
   (b) 1 amu and +1
   (c) 1 amu and 0
   (d) 0 amu and −1
   (e) 0 amu and +1

8. What is the approximate mass and relative charge of a positron?
   (a) 4 amu and +2
   (b) 1 amu and +1
   (c) 1 amu and 0
   (d) 0 amu and −1
   (e) 0 amu and +1

9. What is the approximate mass and relative charge of a proton?
   (a) 4 amu and +2
   (b) 1 amu and +1
   (c) 1 amu and 0
   (d) 0 amu and −1
   (e) 0 amu and +1

10. What nuclide is produced when a K-43 nucleus decays by beta emission?

    (a) ${}^{43}_{18}Ar$ (b) ${}^{42}_{19}K$ (c) ${}^{42}_{20}Ca$ (d) ${}^{43}_{20}Ca$ (e) ${}^{44}_{20}Ca$

11. What nuclide is produced when an O-15 nucleus decays by positron emission?

    (a) ${}^{14}_{6}C$ (b) ${}^{14}_{7}N$ (c) ${}^{15}_{7}N$ (d) ${}^{14}_{8}O$ (e) ${}^{15}_{9}F$

12. What nuclide is produced when a Ba-133 nucleus decays by electron capture?

    (a) ${}^{132}_{55}Cs$ (b) ${}^{133}_{55}Cs$ (c) ${}^{132}_{56}Ba$ (d) ${}^{132}_{57}La$ (e) ${}^{133}_{57}La$

Section 18.3 *Radioactive Decay Series*

13. The uranium-238 decay series begins with the emission of an alpha particle. If the daughter decays by beta emission, what is the resulting particle?
    (a) $^{234}_{90}Th$
    (b) $^{233}_{91}Pa$
    (c) $^{234}_{91}Pa$
    (d) $^{234}_{93}Np$
    (e) none of the above

14. In the final step of the uranium-238 disintegration series, the parent nuclide decays into lead-206 and an alpha particle. What is the parent nuclide?
    (a) $^{202}_{80}Hg$
    (b) $^{210}_{83}Bi$
    (c) $^{206}_{84}Po$
    (d) $^{210}_{84}Po$
    (e) none of the above

Section 18.4 *Radioactive Half-Life*

15. Each "click" registered by a Geiger counter indicates which of the following?
    (a) one half-life
    (b) one scintillation
    (c) one minute elapsing
    (d) one second elapsing
    (e) one nucleus decaying

16. The initial activity of a radionuclide was 1000 dpm and dropped to 125 dpm after 72 days. What is the half-life of the radionuclide?
    (a) $t_{1/2}$ = 18 days
    (b) $t_{1/2}$ = 24 days
    (c) $t_{1/2}$ = 36 days
    (d) $t_{1/2}$ = 144 days
    (e) $t_{1/2}$ = 216 days

17. If 100 mg of a radionuclide is used for medical diagnosis, how much of the radionuclide is still radioactive after 30 hours? *($t_{1/2}$ = 15 hours)*
    (a) 25 mg
    (b) 33 mg
    (c) 50 mg
    (d) 100 mg
    (e) 300 mg

18. If the initial activity of a radionuclide is 160 dpm, how much time elapses before the activity drops to 10 dpm? *($t_{1/2}$ = 48 minutes)*
    (a) 12 min
    (b) 96 min
    (c) 144 min
    (d) 192 min
    (e) 768 min

Section 18.5 *Applications of Radionuclides*

19. An archaeologist discovers an artifact with a carbon-14 activity of 7.6 dpm per gram of carbon. If carbon-14 has an activity of 15.3 dpm, what is the estimated age of the artifact? *($t_{1/2}$ = 5730 years)*
    (a) 1400 years
    (b) 2900 years
    (c) 5700 years
    (d) 11,000 years
    (e) 17,000 years

20. What radionuclide can be used to irradiate food and destroy microorganisms?
    (a) cobalt-60
    (b) iodine-131
    (c) iridium-192
    (d) sodium-24
    (e) xenon-133

Section 18.6 *Induced Radioactivity*

21. Bombarding sodium-23 with a proton produces particle X and a neutron. What is particle X ?
    (a) neon-23
    (b) sodium-24
    (c) magnesium-23
    (d) magnesium-24
    (e) none of the above

22. Firing an accelerated particle at a boron-10 target nucleus produces nitrogen-14 and a gamma ray. What is the projectile particle?
    (a) alpha
    (b) beta
    (c) neutron
    (d) proton
    (e) positron

Section 18.7 *Nuclear Fission*

23. A single neutron causes uranium-235 to fission and release three neutrons. Assuming each of the neutrons causes a second fission that releases two neutrons, how many neutrons are released in the second step?
    (a) 3
    (b) 6
    (c) 8
    (d) 9
    (e) none of the above

24. How many neutrons are produced from the following fission reaction?

$$^{235}_{92}\text{U} + {}^{1}_{0}\text{n} \rightarrow {}^{137}_{52}\text{Te} + {}^{96}_{40}\text{Zr} + ?\,{}^{1}_{0}\text{n}$$

    (a) 1
    (b) 2
    (c) 3
    (d) 4
    (e) none of the above

Section 18.8 *Nuclear Fusion*

25. Which of the following best describes the process of nuclear fusion?
    (a) Combining two light nuclei into a heavy nucleus at 25 °C.
    (b) Combining two light nuclei into a heavy nucleus at ~10,000,000 °C.
    (c) Combining two light nuclei into a heavy nucleus at ~1000 K.
    (d) Splitting a heavy nucleus into two lighter nuclei at 25 °C.
    (e) Splitting a heavy nucleus into two lighter nuclei at ~10,000,000 °C.

26. The nuclear fusion of deuterium and tritium gives a neutron and particle-X. Identify particle-X.
    (a) hydrogen-3
    (b) helium-3
    (c) helium-4
    (d) lithium-5
    (e) none of the above

Nuclear Equations for Transmutation Reactions

27. Phosphorus-30 was the first synthetic radionuclide and was produced by bombarding aluminum-27 with an alpha particle. Write a balanced nuclear equation for the reaction.

28. Element 106 was first synthesized in Germany by firing an oxygen-18 projectile at a californium-249 target. Write a balanced nuclear equation for the reaction if the nuclide produced had a mass number of 263.

Nuclear Equations for Fission and Fusion Reactions

29. A neutron can produce a fission reaction of uranium-235 to give xenon-142, strontium-91, and neutrons. Write a balanced nuclear equation for the reaction.

30. Deuterium and tritium undergo a fusion reaction to produce helium-4 and a neutron at 40,000,000 K. Write a balanced nuclear equation for the reaction.

Chapter 18 Key Terms

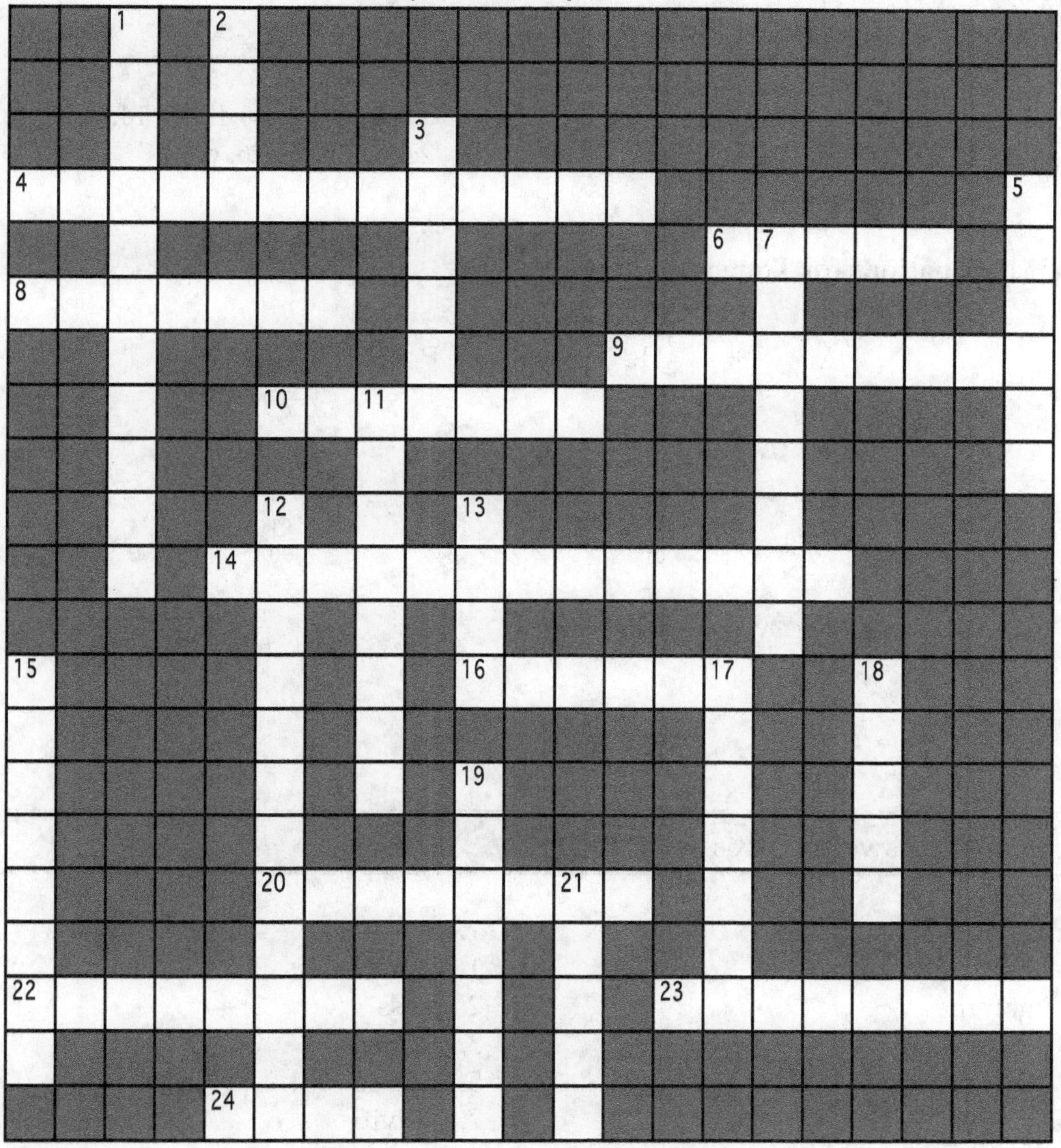

Across

4. the conversion of one element into another
6. a nuclear radiation identical to high-energy light
8. a reaction in which a nucleus splits
9. the nuclide of hydrogen having one neutron
10. a heavy nuclide seizes a core electron
14. the emission of particles from an unstable nucleus
16. a reaction in which two light nuclei combine
20. the minimum mass that sustains a chain reaction
22. using atomic notation to describe a reaction
23. a description of a change involving a nucleus
24. a nuclear radiation identical to an electron

Down

1. stepwise disintegration of a radionuclide (2 words)
2. the *A* value refers to the ________ number
3. a nuclide that decays to give a daughter product
5. the *Z* value refers to the ________ number
7. the rate of decay of radionuclides
11. a positive electron
12. a nucleus of a specific radioactive isotope
13. the life of 50% of the radionuclides in a sample
15. the nuclide resulting from a decaying nucleus
17. a nucleus of a specific isotope
18. a fission reaction that initiates additional fissions
19. the nuclide of hydrogen having two neutrons
21. a nuclear radiation identical to a helium-4 nucleus

Chapter 18 Answers to Self-Test

1. **e** 2. **a** 3. **c** 4. **a** 5. **c** 6. **c** 7. **d** 8. **e** 9. **b** 10. **d** 11. **c** 12. **b** 13. **c** 14. **d** 15. **e**
16. **b** 17. **a** 18. **d** 19. **c** 20. **a** 21. **c** 22. **a** 23. **b** 24. **c** 25. **b** 26. **c**

Nuclear Equations for Transmutation Reactions

27. $^{27}_{13}\text{Al} + ^{4}_{2}\text{He} \rightarrow ^{30}_{15}\text{P} + ^{1}_{0}\text{n}$

28. $^{249}_{98}\text{Cf} + ^{18}_{8}\text{O} \rightarrow ^{263}_{106}\text{Sg} + 4\,^{1}_{0}\text{n}$

29. $^{235}_{92}\text{U} + ^{1}_{0}\text{n} \rightarrow ^{142}_{54}\text{Xe} + ^{91}_{38}\text{Sr} + 3\,^{1}_{0}\text{n}$

30. $^{2}_{1}\text{H} + ^{3}_{1}\text{H} \rightarrow ^{4}_{2}\text{He} + ^{1}_{0}\text{n}$

Chapter 18 Answers to Crossword Puzzle

Across

4. transmutation
6. gamma
8. fission
9. deuterium
10. capture
14. radioactivity
16. fusion
20. critical
22. equation
23. reaction
24. beta

Down

1. decay series
2. mass
3. parent
5. atomic
7. activity
11. positron
12. radionuclide
13. half
15. daughter
17. nuclide
18. chain
19. tritium
21. alpha

CHAPTER

19

Organic Chemistry

Section 19.1 *Hydrocarbons*

1. Which of the following is a class of saturated hydrocarbons?
   (a) alkanes
   (b) alkenes
   (c) alkynes
   (d) aromatic
   (e) none of the above

Section 19.2 *Alkanes*

2. What is the IUPAC name of the following alkane:

   CH_3-CH_2-CH_2-CH_2-CH_2-CH_2-CH_2-CH_2-CH_3?

   (a) decane
   (b) heptane
   (c) hexane
   (d) nonane
   (e) none of the above

3. How many isomers share the molecular formula C_6H_{14}?
   (a) 2
   (b) 3
   (c) 4
   (d) 5
   (e) 6

4. What is the name of the following alkyl group: $(CH_3)_2CH$–?
   (a) butyl
   (b) ethyl
   (c) isopropyl
   (d) methyl
   (e) propyl

5. Which of the following is the structural formula for 2-methylhexane?
   (a) CH_3-$CH(CH_3)$-CH_2-CH_2-CH_3
   (b) CH_3-$CH(CH_3)$-CH_2-CH_2-CH_2-CH_3
   (c) CH_3-CH_2-$CH(CH_3)$-CH_2-CH_2-CH_3
   (d) CH_3-$CH(CH_3)$-CH_2-CH_2-CH_2-CH_2-CH_3
   (e) CH_3-CH_2-$CH(CH_3)$-CH_2-CH_2-CH_2-CH_3

6. Which of the following is a balanced equation for the complete combustion of propane?
   (a) $2\,C_3H_8 + 3\,O_2 \xrightarrow{\text{spark}} 6\,CO + 8\,H_2$
   (b) $2\,C_3H_8 + 7\,O_2 \xrightarrow{\text{spark}} 6\,CO + 8\,H_2O$
   (c) $C_3H_8 + 3\,O_2 \xrightarrow{\text{spark}} 3\,CO_2 + 4\,H_2$
   (d) $C_3H_8 + 5\,O_2 \xrightarrow{\text{spark}} 3\,CO_2 + 4\,H_2O$
   (e) $C_3H_8 + O_2 \xrightarrow{\text{spark}}$ no reaction

Section 19.3 *Alkenes and Alkynes*

7. Which of the following is the condensed structural formula for 3-octene?
   (a) CH_3-CH_2-CH=CH-CH_2-CH_2-CH_3
   (b) CH_3-CH_2-CH_2-CH=CH-CH_2-CH_3
   (c) CH_3-CH_2-CH_2-CH_2-CH=CH-CH_2-CH_3
   (d) CH_3-CH_2-CH_2-CH=CH-CH_2-CH_2-CH_3
   (e) none of the above

8. What is the IUPAC name of the following: CH_3-CH_2-C≡CH?
   (a) butyne
   (b) 1-butyne
   (c) 2-butyne
   (d) 3-butyne
   (e) none of the above

9. What are the products from the complete combustion of acetylene?
   (a) C_2H_2 and H_2O
   (b) CO and H_2O
   (c) CO and H_2
   (d) CO_2 and H_2O
   (e) CO_2 and H_2

10. Which of the following illustrates the hydrogenation of propylene gas with nickel catalyst at 25 °C and 1 atm?

(a) $CH_2{=}CH{-}CH_3 + H_2 \xrightarrow{Ni} C_3H_4$

(b) $CH_2{=}CH{-}CH_3 + H_2 \xrightarrow{Ni} C_3H_6$

(c) $CH_2{=}CH{-}CH_3 + H_2 \xrightarrow{Ni} C_3H_8$

(d) $CH_2{=}CH{-}CH_3 + H_2 \xrightarrow{Ni} C_6H_6$

(e) $CH_2{=}CH{-}CH_3 + H_2 \xrightarrow{Ni}$ no reaction

Section 19.4 *Arenes*

11. How many representations of benzene are there according to the delocalized electron model?
   (a) 1
   (b) 2
   (c) 3
   (d) 4
   (e) none of the above

12. How many different isomers are there for dimethylbenzene, $C_6H_4(CH_3)_2$?
   (a) 0
   (b) 2
   (c) 3
   (d) 4
   (e) none of the above

Section 19.5 *Hydrocarbon Derivatives*

13. Which of the following classes of compounds does *not* have a carbonyl group?
   (a) aldehyde
   (b) amide
   (c) ester
   (d) ketone
   (e) phenol

14. What is the general formula for an organic halide?
   (a) R–X
   (b) R–OH
   (c) Ar–OH
   (d) R–O–R′
   (e) $R{-}NH_2$

15. What is the general formula for an alcohol?
   (a) R-X
   (b) R-OH
   (c) Ar-OH
   (d) R-O-R′
   (e) $R\text{-}NH_2$

16. What is the general formula for an ether?
   (a) R-X
   (b) R-OH
   (c) Ar-OH
   (d) R-O-R′
   (e) $R\text{-}NH_2$

17. What class of compound has the following general formula: $R-\overset{\overset{\displaystyle O}{\|}}{C}-H$?
   (a) aldehyde
   (b) ketone
   (c) carboxylic acid
   (d) ester
   (e) amide

18. What class of compound has the following general formula: $R-\overset{\overset{\displaystyle O}{\|}}{C}-OH$?
   (a) aldehyde
   (b) ketone
   (c) carboxylic acid
   (d) ester
   (e) amide

19. What class of compound is the following: $C_6H_5\text{-}OH$?
   (a) organic halide
   (b) alcohol
   (c) phenol
   (d) ether
   (e) amine

20. What class of compound is the following: $CH_3CH_2\text{-}NH_2$?
   (a) organic halide
   (b) alcohol
   (c) phenol
   (d) ether
   (e) amine

Section 19.6 *Organic Halides*

21. What is the condensed structural formula for ethyl bromide?
    (a) CH_3-Br
    (b) CH_3-CH_2-Br
    (c) CH_3-CH_2-CH_2-Br
    (d) CH_3-CHBr-CH_3
    (e) none of the above

Section 19.7 *Alcohols, Phenols, and Ethers*

22. What is the condensed structural formula for isopropyl alcohol?
    (a) CH_3-CH_2-OH
    (b) CH_3-CH_2-CH_2-OH
    (c) CH_3-CH(OH)-CH_3
    (d) CH_3-CH(OH)-CH_2-CH_3
    (e) none of the above

23. What is the common name for the following: CH_3CH_2-O-CH_3?
    (a) diethyl ether
    (b) ethyl ether
    (c) ethyl methyl ether
    (d) methyl ether
    (e) none of the above

Section 19.8 *Amines*

24. What is the condensed structural formula for propylamine?
    (a) CH_3-CH_2-NH_2
    (b) CH_3-CH_2-CH_2-NH_2
    (c) CH_3-CH(NH_2)-CH_3
    (d) CH_3-CH(NH_2)-CH_2-CH_3
    (e) none of the above

Section 19.9 *Aldehydes and Ketones*

25. Which of the following organic compounds is most likely an aldehyde based upon the nomenclature suffix?
    (a) chlordane (insecticide)
    (b) cholesterol (steroid)
    (c) citral (lemon fragrance)
    (d) codeine (narcotic)
    (e) cresol (antiseptic)

26. What is the IUPAC systematic name for the following: $CH_3-\overset{\overset{\displaystyle O}{\|}}{C}-CH_3$?

(a) acetone
(b) dimethyl ketone
(c) propanaldehyde
(d) propanal
(e) propanone

Section 19.10 *Carboxylic Acids, Esters, and Amides*

27. What is the IUPAC systematic name for the following: $CH_3-\overset{\overset{\displaystyle O}{\|}}{C}-OH$?

(a) acetic acid
(b) ethanoic acid
(c) formic acid
(d) methanoic acid
(e) propanoic acid

28. What is the IUPAC systematic name for the following: $H-\overset{\overset{\displaystyle O}{\|}}{C}-O-CH_2CH_3$?

(a) ethyl formate
(b) ethyl methanoate
(c) ethyl propanoate
(d) methyl acetate
(e) methyl ethanoate

IUPAC Nomenclature of Organic Compounds

29. What is the IUPAC name for the following compound?

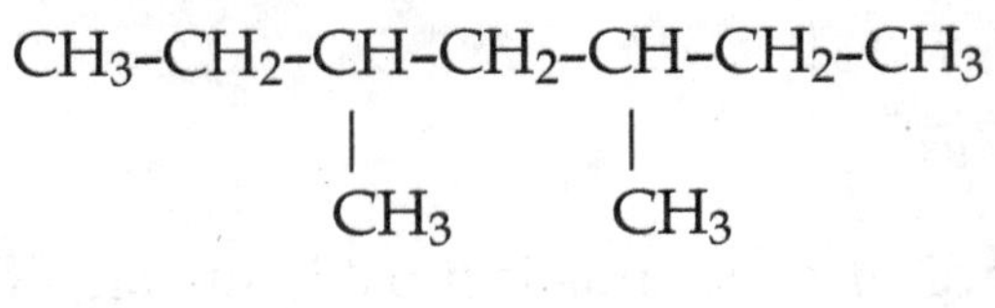

30. What is the IUPAC name for the following compound?

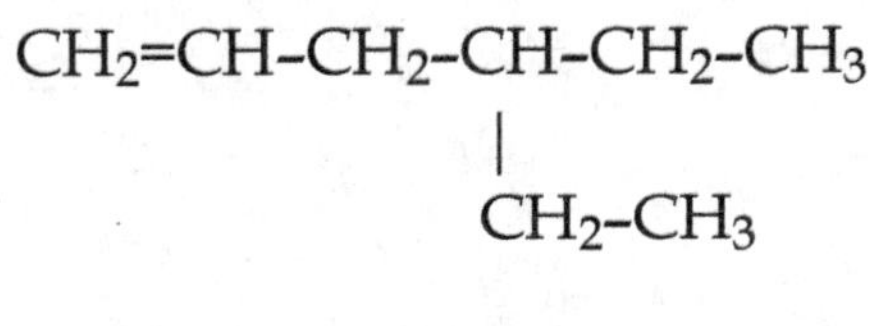

Chapter 19 Key Terms

Across

1. a family of saturated hydrocarbons
3. a compound containing C, H, and O
6. a hydrocarbon containing all single bonds
10. a carbon and oxygen double bond
11. the second member of the alkane family
12. a compound containing only H and C
17. same molecular formula but different structures
19. a formula showing the molecular arrangement
20. an alkane fragment after removing a H atom
21. a family of compounds
22. a small molecule that repeats in a polymer

Down

1. a reaction of a hydrocarbon with H_2 or Br_2
2. a group typical of a class of compounds
4. a family of aromatic hydrocarbons
5. a long-chain compound of joined small molecules
7. an arene fragment after removing a H atom
8. the functional group in a carboxylic acid
9. a benzene fragment after removing a H atom
10. a reaction of a hydrocarbon with O_2
12. the functional group in an alcohol or phenol
13. the compounds containing carbon
14. a family of compounds with a double bond
15. a hydrocarbon containing a double or triple bond
16. a hydrocarbon containing a benzene ring
18. a family of compounds with a triple bond

Chapter 19 Answers to Self-Test

1. a 2. d 3. d 4. c 5. b 6. d 7. c 8. b 9. d 10. c 11. a 12. c 13. e 14. a 15. b
16. d 17. a 18. c 19. c 20. e 21. b 22. c 23. c 24. b 25. c 26. e 27. b 28. b

IUPAC Nomenclature of Organic Compounds

29. 3,5-dimethylheptane

30. 4-ethyl-1-hexene

Chapter 19 Answers to Crossword Puzzle

Across

1. alkanes
3. derivative
6. saturated
10. carbonyl
11. ethane
12. hydrocarbon
17. isomers
19. structural
20. alkyl
21. class
22. monomer

Down

1. addition
2. functional
4. arenes
5. polymer
7. aryl
8. carboxyl
9. phenyl
10. combustion
12. hydroxyl
13. organic
14. alkenes
15. unsaturated
16. aromatic
18. alkynes

CHAPTER

20

Biochemistry

Section 20.1 *Biological Compounds*

1. What type of biological compound is a polymer of amino acids?
   (a) protein
   (b) carbohydrate
   (c) lipid
   (d) nucleic acid
   (e) none of the above

2. What type of biological compound is a polymer of simple sugar molecules?
   (a) protein
   (b) carbohydrate
   (c) lipid
   (d) nucleic acid
   (e) none of the above

3. What type of biological compound is characterized by carboxylic acid, alcohol, and ester functional groups?
   (a) protein
   (b) carbohydrate
   (c) lipid
   (d) nucleic acid
   (e) none of the above

4. What type of biological compound is characterized by aldehyde, alcohol, and amine functional groups?
   (a) protein
   (b) carbohydrate
   (c) lipid
   (d) nucleic acid
   (e) none of the above

5. Which of the following type of bonds is found in a protein?
   (a) peptide linkage
   (b) glycoside linkage
   (c) ester linkage
   (d) phosphate linkage
   (e) none of the above

6. Which of the following type of bonds is found in a carbohydrate?
   (a) peptide linkage
   (b) glycoside linkage
   (c) ester linkage
   (d) phosphate linkage
   (e) none of the above

7. Which of the following type of bonds is found in a lipid?
   (a) peptide linkage
   (b) glycoside linkage
   (c) ester linkage
   (d) phosphate linkage
   (e) none of the above

8. Which of the following type of bonds is found in a nucleic acid?
   (a) peptide linkage
   (b) glycoside linkage
   (c) ester linkage
   (d) phosphate linkage
   (e) none of the above

Section 20.2 *Proteins*

9. What is the overall shape of a protein that serves a structural function, such as movement or locomotion?
   (a) compact and spherical
   (b) long and extended
   (c) random shape
   (d) all of the above
   (e) none of the above

10. What is the overall shape of a protein that serves a metabolic role, such as oxygen transport?
    (a) compact and spherical
    (b) long and extended
    (c) random shape
    (d) all of the above
    (e) none of the above

11. What is the overall shape of a protein after it undergoes denaturation?
   (a) compact and spherical
   (b) long and extended
   (c) random shape
   (d) all of the above
   (e) none of the above

12. What type of bonds are responsible for the secondary structure of a protein?
   (a) amide bonds
   (b) ester bonds
   (c) hydrogen bonds
   (d) ionic bonds
   (e) none of the above

13. What type of protein structure corresponds to a pleated sheet of amino acids?
   (a) primary
   (b) secondary
   (c) tertiary
   (d) quaternary
   (e) none of the above

14. How many amino acids compose valylalanyltyrosine?
   (a) 1
   (b) 2
   (c) 3
   (d) 4
   (e) 1000s

15. How many amino acid sequences are possible for a tripeptide containing proline (Pro), histidine (His), and leucine (Leu)?
   (a) 1
   (b) 3
   (c) 6
   (d) 9
   (e) 1000s

Section 20.3 *Enzymes*

16. An enzyme is an example of which type of biological compound?
   (a) protein
   (b) carbohydrate
   (c) lipid
   (d) nucleic acid
   (e) none of the above

17. Which of the following occurs in Step 1 of enzyme catalysis?
    (a) The substrate cleaves the enzyme molecule.
    (b) The enzyme cleaves the substrate molecule.
    (c) The substrate molecule binds to the active site.
    (d) The enzyme molecule binds to the active site.
    (e) none of the above

18. Which of the following occurs in Step 2 of enzyme catalysis?
    (a) The substrate cleaves the enzyme molecule.
    (b) The enzyme cleaves the substrate molecule.
    (c) The substrate molecule binds to the active site.
    (d) The enzyme molecule binds to the active site.
    (e) none of the above

19. What is the term for a molecule that blocks the active site on an enzyme?
    (a) catalyst
    (b) inhibitor
    (c) steroid
    (d) substrate
    (e) none of the above

Section 20.4 *Carbohydrates*

20. Which of the following functional groups are found in a ketopentose?
    (a) alcohol and aldehyde
    (b) alcohol and ketone
    (c) aldehyde and ketone
    (d) aldehyde and phenol
    (e) none of the above

21. How many carbon atoms are in a molecule of ketopentose sugar?
    (a) 1
    (b) 5
    (c) 6
    (d) 12
    (e) none of the above

22. Which monosaccharides are produced from the acid hydrolysis of sucrose?
    (a) fructose and glucose
    (b) galactose and glucose
    (c) gulose and glucose
    (d) ribose and glucose
    (e) none of the above

23. What is the repeating monosaccharide in cellulose?
  (a) fructose
  (b) galactose
  (c) glucose
  (d) ribose
  (e) none of the above

Section 20.5 *Lipids*

24. Which of the following are characteristic of a lipid fat?
  (a) liquid
  (b) plant source
  (c) unsaturated fatty acids
  (d) water-insoluble
  (e) all of the above

25. Which of the following is *not* found in a fat or oil?
  (a) ester linkages
  (b) glycerol
  (c) saturated fatty acids
  (d) steroid
  (e) unsaturated fatty acids

26. Which of the following is the structure of palmitic acid?
  (a) CH_3-$(CH_2)_{10}$-COOH
  (b) CH_3-$(CH_2)_{12}$-COOH
  (c) CH_3-$(CH_2)_{14}$-COOH
  (d) CH_3-$(CH_2)_{16}$-COOH
  (e) none of the above

27. What are the products from the saponification of trilinolein?
  (a) water and sodium linoleate
  (b) water and linoleic acid
  (c) glycerol and linoleic acid
  (d) glycerol and sodium linoleate
  (e) none of the above

28. What are the products from the saponification of a lipid wax?
  (a) long-chain alcohol and a fatty acid
  (b) long-chain alcohol and a sodium salt of a fatty acid
  (c) glycerol and a fatty acid
  (d) glycerol and a sodium salt of a fatty acid
  (e) none of the above

Section 20.6 *Nucleic Acids*

29. Which of the following is found in both DNA and RNA nucleotides?
    (a) deoxyribose sugar
    (b) phosphoric acid
    (c) ribose sugar
    (d) thymine
    (e) uracil

30. How many nucleotide sequences are possible for a DNA trinucleotide having a cytosine base and two guanine bases?
    (a) 1
    (b) 3
    (c) 6
    (d) 9
    (e) none of the above

31. During DNA replication, an adenine base on the template DNA strand will code for which base in the complementary DNA strand?
    (a) cytosine
    (b) guanine
    (c) thymine
    (d) uracil
    (e) the base varies

32. During RNA transcription, an adenine base on the template DNA strand will code for which base on the growing RNA strand?
    (a) cytosine
    (b) guanine
    (c) thymine
    (d) uracil
    (e) the base varies

Protein Synthesis

33. Explain the process of protein synthesis, and the role of DNA and RNA. Define the term *codon*.

Chapter 20 Key Terms

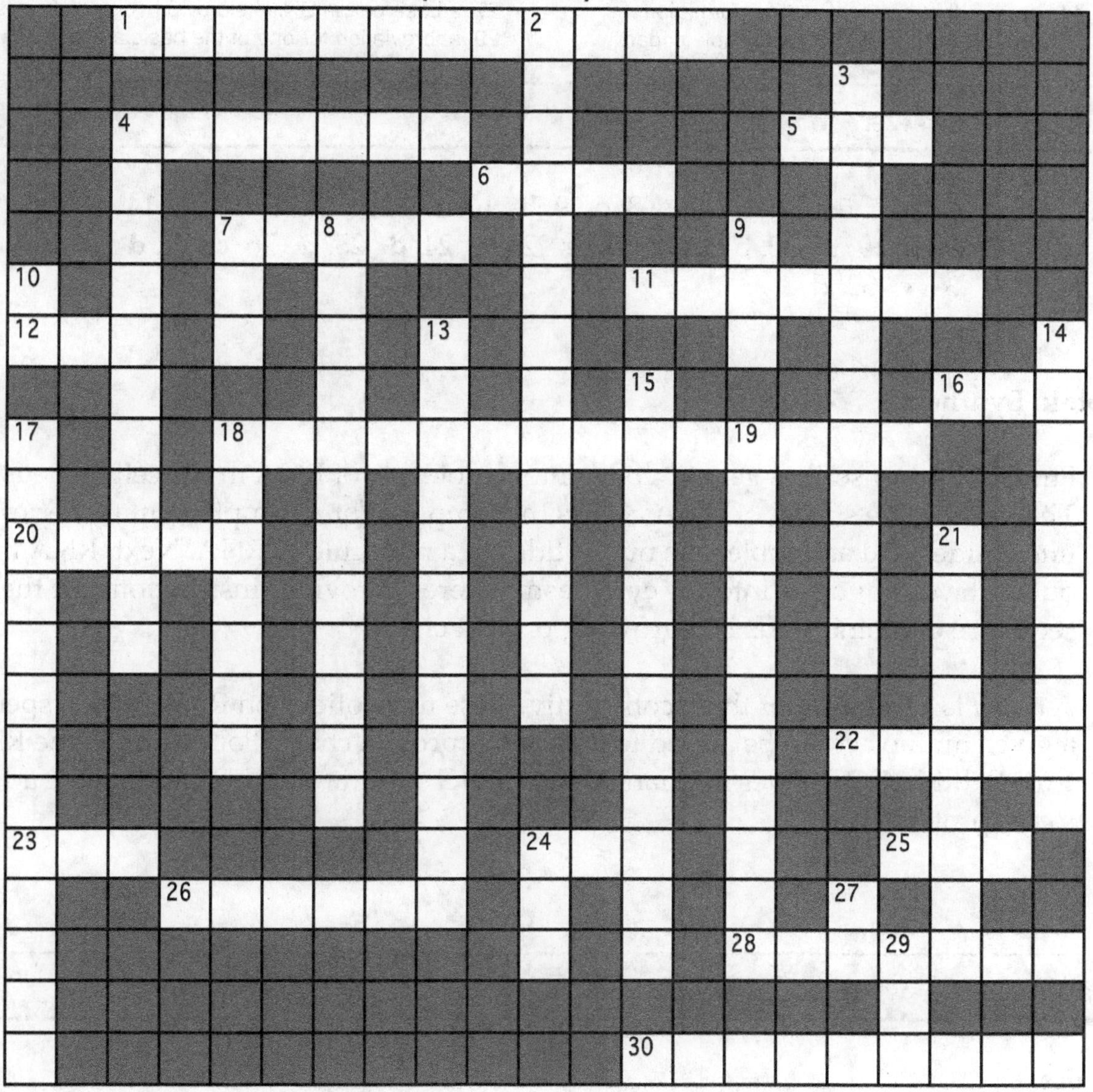

Across

1. the study of biological compounds and reactions
4. a type of bond that joins two amino acids
5. an ester of a long-chain alcohol and a fatty acid
6. symbolizes the products in the enzyme model
7. an acid with a long hydrocarbon side-chain
11. a lipid containing four rings of carbon atoms
12. abbreviation for the amino acid tryptophan
16. abbreviation for the amino acid leucine
18. a carbohydrate polymer of simple sugars
20. a carboxylic acid containing an $-NH_2$ group
22. a water-insoluble compound such as a fat or oil
23. abbreviation for the amino acid phenylalanine
24. abbreviation for one of the acidic amino acids
25. a lipid that contains mostly unsaturated fatty acids

Down

2. a reaction of a triglyceride and sodium hydroxide
3. a polyhydroxy aldehyde or ketone
4. ~50 amino acids in a long-chain molecule
7. a lipid that contains mostly saturated fatty acids
8. an ester of glycerol and three fatty acids
9. symbolizes the enzyme in the enzyme model
10. a base pair in DNA held by 2 hydrogen bonds
13. a carbohydrate composed of two simple sugars
14. a repeating unit in a nucleic acid
15. an ester of glycerol, 2 fatty acids, and H_3PO_4
17. the process by which DNA synthesizes RNA
18. a polymer of amino acids
19. the process by which DNA synthesizes DNA
21. two amino acids joined by an amide bond

26. a catalyst for a biochemical reaction
28. an acid that carries genetic information
30. an –O– bond that joins two simple sugars

24. abbreviation for the amino acid alanine
27. a base pair in DNA held by 3 hydrogen bonds
29. abbreviation for one of the basic amino acids

Chapter 20 Answers to Self-Test

1. a 2. b 3. c 4. d 5. a 6. b 7. c 8. d 9. b 10. a 11. c 12. c 13. b 14. c 15. c 16. a 17. c 18. b 19. b 20. b 21. b 22. a 23. c 24. d 25. d 26. c 27. d 28. b 29. b 30. b 31. c 32. d

Protein Synthesis

33. First, DNA transcribes genetic code onto a molecule of RNA in the cell nucleus. That is, a single strand of DNA acts as the template for a complementary series of nucleotides and assembles the nucleotides in a molecule of RNA. Next, RNA moves out of the cell nucleus into the cytoplasm where it provides instructions for the sequence of amino acids in a growing protein chain.

    A *codon* is composed of three consecutive base nucleotides on RNA, which specifies a given amino acid to be added to a growing protein chain. For example, the RNA trinucleotide GCG codes for alanine, and AUG specifies methionine is to be added to the protein chain.

Chapter 20 Answers to Crossword Puzzle

Across
1. biochemistry
4. peptide
5. wax
6. lock
7. fatty
11. steroid
12. Trp
16. Leu
18. polysaccharide
20. amino
22. lipid
23. Phe
24. Asp
25. oil
26. enzyme
28. nucleic
30. glycoside

Down
2. saponification
3. carbohydrate
4. polypeptide
7. fat
8. triglyceride
9. key
10. AT (A=T)
13. disaccharide
14. nucleotide
15. phospholipid
17. transcription
18. protein
19. replication
21. dipeptide
24. Ala
27. GC (G≡C)
29. Lys

Selected Solutions Manual

for

Odd-Numbered Exercises

Introductory Chemistry
CONCEPTS AND CRITICAL THINKING

EIGHTH EDITION

CHAPTER

1

Introduction to Chemistry

Section 1.1 *Evolution of Chemistry*

1. The Chinese believed that two forces, *yin* and *yang*, were responsible for bringing the earthly world and everything in nature into existence.

3. Empedocles stated *air, earth, fire,* and *water* were four basic elements responsible for the natural world.

5. Step 1 of the scientific method is to *plan and perform an experiment, make observations,* and *record data.*

7. Step 3 of the scientific method is to *conduct additional experiments* in order to test the hypothesis.

9. A *scientific theory* explains the behavior of nature using a model or concept, whereas a *natural law* states a relationship that is measurable under various experimental conditions and can be stated in the form of an equation.

11. (a) is a *natural law* because mass can be measured.
    (b) is a *scientific theory* because gas molecule collisions cannot be measured.
    (c) is a *natural law* because gas volumes can be measured.
    (d) is a *scientific theory* because it is a model that explains the nucleus.

Section 1.2 *Modern Chemistry*

13. *Chemistry* is referred to as the central science because it supports other physical sciences, life sciences, and earth sciences. An understanding of chemistry is also prerequisite to the health sciences.

15. Scientists and engineers must have formal training in chemistry, as well as doctors, nurses, dentists, physical therapists, chiropractors, dietitians, and veterinarians.

Section 1.3 *Learning Chemistry*

17. A solution to the nine-dot problem with only one straight line is to use a *very wide* line; the unconscious assumption regards the thickness of the line.

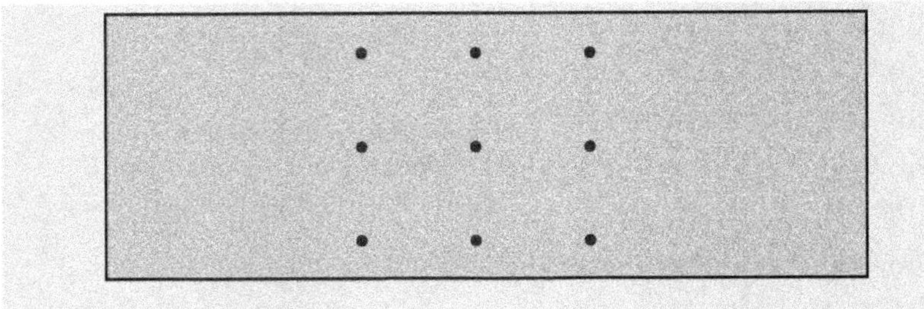

A more creative solution using one straight line is as follows. Draw nine dots on a piece of paper and curl the page into a roll. Draw a line through the upper three dots and continue drawing the line around the page until all nine dots are connected. (Alternatively, wrap a piece of thread continuously around the rolled page; the thread will eventually touch all nine dots.) The unconscious assumption regards a line that is straight in one dimension, but curved in another.

Challenge Exercise

19. Although the perception is that line segment AB is shorter than BC, the two lines are of equal length. Measure each line segment to confirm this statement.

CHAPTER

PSS

Prerequisite Science Skills

Section PSS.1 *Measurements*

1. The mass of a 5¢ nickel coin is approximately 5 g.

3.

	Unit	Quantity		Unit	Quantity
(a)	centimeter	length	(b)	gram	mass
(c)	milliliter	volume	(d)	second	time

5. (a) 5.0 cm and (d) 0.5 cm each have an uncertainty of ± 0.1 cm.

7. (d) 75.518 g is the only mass with an uncertainty of ± 0.001 g.

9. (b) 32.00 mL and (c) 32.05 mL each have an uncertainty of ± 0.05 mL.

Section PSS.2 *Significant Digits*

11.

	Measurement	Significant Digits
(a)	20 cm	1 significant digit
(b)	200 g	1 significant digit
(c)	0.002 mL	1 significant digit
(d)	0.0002 s	1 significant digit

13.

	Measurement	Significant Digits
(a)	0.5 cm	1 significant digit
(b)	1055 g	4 significant digits
(c)	1.2×10^{0} mL	2 significant digits
(d)	4.50×10^{1} s	3 significant digits

15.

	Measurement	Significant Digits
(a)	5.00 cm	3 significant digits
(b)	5005 g	4 significant digits
(c)	1×10^{-1} mL	1 significant digit
(d)	1.0×10^{-2} s	2 significant digits

Section PSS.3 *Rounding Off Nonsignificant Digits*

17.

	Example	Rounded Off
(a)	10.25	10.3
(b)	10.20	10.2
(c)	0.01029	0.0103
(d)	10,248	10,200

19.

	Example	Rounded Off
(a)	9.123×10^{5}	9.12×10^{5}
(b)	9.456×10^{10}	9.46×10^{10}
(c)	9.000×10^{-7}	9.00×10^{-7}
(d)	9.075×10^{-12}	9.08×10^{-12}

Section PSS.4 *Adding and Subtracting Measurements*

21. (a) 1.55 cm + 36.15 cm + 17.3 cm = 55.00 cm *rounds to* 55.0 cm
    (b) 5.0 cm + 16.3 cm + 0.95 cm = 22.25 cm *rounds to* 22.3 cm

23. (a) 22.10 cm – 10.5 cm = 11.60 cm *rounds to* 11.6 cm
    (b) 10.0 cm – 0.15 cm = 9.85 cm *rounds to* 9.9 cm

Section PSS.5 *Multiplying and Dividing Measurements*

25. (a) $1.25 \text{ cm} \times 0.5 \text{ cm} = 0.625 \text{ cm}^2$ *rounds to* 0.6 cm^2
    (b) $2.55 \text{ cm} \times 1.1 \text{ cm} = 2.805 \text{ cm}^2$ *rounds to* 2.8 cm^2
    (c) $12.0 \text{ cm}^2 \times 1.00 \text{ cm} = 12.0 \text{ cm}^3$ *rounds to* 12.0 cm^3
    (d) $22.1 \text{ cm}^2 \times 0.75 \text{ cm} = 16.575 \text{ cm}^3$ *rounds to* 17 cm^3

27. (a) $\frac{66.3 \text{ g}}{7.5 \text{ mL}} = 8.84 \text{ g/mL}$ *rounds to* 8.8 g/mL
    (b) $\frac{12.5 \text{ g}}{4.1 \text{ mL}} = 3.04878 \text{ g/mL}$ *rounds to* 3.0 g/mL
    (c) $\frac{42.620 \text{ g}}{10.0 \text{ mL}} = 4.262 \text{ g/mL}$ *rounds to* 4.26 g/mL
    (d) $\frac{91.235 \text{ g}}{10.00 \text{ mL}} = 9.1235 \text{ g/mL}$ *rounds to* 9.124 g/mL

Section PSS.6 *Exponential Numbers*

29. (a) $10 \times 10 \times 10 = 10^{3}$
    (b) $\frac{1}{10} \times \frac{1}{10} \times \frac{1}{10} = \left(\frac{1}{10}\right)^3 = 10^{-3}$

31. (a) $2 \times 2 \times 2 = 2^3$

(b) $\frac{1}{2} \times \frac{1}{2} \times \frac{1}{2} = \left(\frac{1}{2}\right)^3 = 2^{-3}$

33.

	Scientific Notation	Ordinary Number
(a)	1×10^3	1000
(b)	1×10^{-7}	0.000 000 1

35.

	Ordinary Number	Scientific Notation
(a)	1,000,000,000	1×10^9
(b)	0.000 000 01	1×10^{-8}

37.

	Scientific Notation	Ordinary Number
(a)	1×10^1	10
(b)	1×10^{-1}	0.1

Section PSS.7 *Scientific Notation*

39.

	Ordinary Number	Scientific Notation
(a)	80,916,000	8.0916×10^7
(b)	0.000 000 015	1.5×10^{-8}
(c)	335,600,000,000,000	3.356×10^{14}
(d)	0.000 000 000 000 927	9.27×10^{-13}

41. 2.69×10^{22} oxygen molecules

43. 5.32×10^{-23} g/oxygen molecule

General Exercises

45. 10.0 mL (± 0.1 mL)

47. 186,000 miles per second (1.86×10^5 miles per second)

49. Remaining length: 255 cm – 2(25.0 cm) = 205 cm

51.

	Exponential Number	Scientific Notation
(a)	352×10^4	3.52×10^6
(b)	0.191×10^{-5}	1.91×10^{-6}

Challenge Exercises

53. Mass of a neutron $= 1.6749 \times 10^{-24}$ g

Mass of a proton $= 1.6726 \times 10^{-24}$ g

Total mass: $(1.6749 \times 10^{-24}\text{ g}) + (1.6726 \times 10^{-24}\text{ g}) = 3.3475 \times 10^{-24}$ g

55. 1 metric ton = 2200 lb = 2.200×10^{3} lb

1 English ton = 2000 lb = 2.000×10^{3} lb

Total Mass: 2.200×10^{3} lb + 2.000×10^{3} lb = 4.200×10^{3} lb

CHAPTER

2

The Metric System

Section 2.1 *Basic Units and Symbols*

1.

	Quantity	Basic Unit		Quantity	Basic Unit
(a)	length	meter	(b)	mass	gram
(c)	volume	liter	(d)	time	second

3.

	Unit	Symbol		Unit	Symbol
(a)	exameter	Em	(b)	teragram	Tg
(c)	attoliter	aL	(d)	picosecond	ps

5.

	Symbol	Unit		Symbol	Unit
(a)	km	kilometer	(b)	Mg	megagram
(c)	nL	nanoliter	(d)	ds	decisecond

7.

	Instrument	Quantity		Instrument	Quantity
(a)	metric ruler	length	(b)	platform balance	mass
(c)	graduated cylinder	volume	(d)	stopwatch	time

9.

	Quantity	Unit/Symbol		Quantity	Unit/Symbol
(a)	1×10^9 m	gigameter, Gm	(b)	1×10^3 g	kilogram, kg
(c)	1×10^{-3} L	milliliter, mL	(d)	1×10^{-9} s	nanosecond, ns

Section 2.2 *Metric Conversion Factors*

11.

	Unit Equation		Unit Equation
(a)	1 km = 1×10^3 m	(b)	1 Mg = 1×10^6 g
(c)	1 L = 1×10^3 mL	(d)	1 s = 1×10^6 μs

13. Unit Factors

(a) $\frac{1\ \text{km}}{1 \times 10^3\ \text{m}}$ and $\frac{1 \times 10^3\ \text{m}}{1\ \text{km}}$

(b) $\frac{1\ \text{Mg}}{1 \times 10^6\ \text{g}}$ and $\frac{1 \times 10^6\ \text{g}}{1\ \text{Mg}}$

(c) $\frac{1\ \text{L}}{1 \times 10^3\ \text{mL}}$ and $\frac{1 \times 10^3\ \text{mL}}{1\ \text{L}}$

(d) $\frac{1\ \text{s}}{1 \times 10^6\ \mu\text{s}}$ and $\frac{1 \times 10^6\ \mu\text{s}}{1\ \text{s}}$

Section 2.3 *Metric–Metric Conversions*

15. (a) $5.00\ \cancel{\text{m}} \times \frac{1\ \text{km}}{1000\ \cancel{\text{m}}} = 5.00 \times 10^{-3}\ \text{km}$

(b) $5.00\ \cancel{\text{g}} \times \frac{100\ \text{cg}}{1\ \cancel{\text{g}}} = 5.00 \times 10^{2}\ \text{cg}$

(c) $5.00\ \cancel{\text{L}} \times \frac{10\ \text{dL}}{1\ \cancel{\text{L}}} = 50.0\ \text{dL}$

(d) $5.00\ \cancel{\text{s}} \times \frac{1 \times 10^9\ \text{ns}}{1\ \cancel{\text{s}}} = 5.00 \times 10^{9}\ \text{ns}$

17. (a) $6.50\ \cancel{\text{Tm}} \times \frac{1 \times 10^{12}\ \cancel{\text{m}}}{1\ \cancel{\text{Tm}}} \times \frac{1\ \text{Mm}}{1 \times 10^6\ \cancel{\text{m}}} = 6.50 \times 10^{6}\ \text{Mm}$

(b) $650\ \cancel{\text{Gg}} \times \frac{1 \times 10^{9}\ \cancel{\text{g}}}{1\ \cancel{\text{Gg}}} \times \frac{1\ \text{kg}}{1000\ \cancel{\text{g}}} = 6.5 \times 10^{8}\ \text{kg}$

(c) $0.650\ \cancel{\text{cL}} \times \frac{1\ \cancel{\text{L}}}{100\ \cancel{\text{cL}}} \times \frac{10\ \text{dL}}{1\ \cancel{\text{L}}} = 6.50 \times 10^{-2}\ \text{dL}$

(d) $0.000\,650\ \cancel{\text{ns}} \times \frac{1\ \cancel{\text{s}}}{1 \times 10^{9}\ \cancel{\text{ns}}} \times \frac{1 \times 10^{12}\ \text{ps}}{1\ \cancel{\text{s}}} = 6.50 \times 10^{-1}\ \text{ps}$ (0.650 ps)

Section 2.4 *Metric–English Conversions*

19. (a) 1 in. = 2.54 cm (b) 1 lb = 454 g
(c) 1 qt = 946 mL (d) 1 sec = 1.00 s

21. (a) $25\ \cancel{\text{cm}} \times \frac{1\ \text{in.}}{2.54\ \cancel{\text{cm}}} = 9.8\ \text{in.}$

(b) $25\ \cancel{\text{g}} \times \frac{1\ \text{lb}}{454\ \cancel{\text{g}}} = 0.055\ \text{lb}$

(c) $2.50\ \cancel{\text{qt}} \times \frac{946\ \text{mL}}{1\ \cancel{\text{qt}}} = 2{,}370\ \text{mL}$

(d) $2.50\ \cancel{\text{sec}} \times \frac{1\ \text{s}}{1.00\ \cancel{\text{sec}}} = 2.50\ \text{s}$

23. (a) $72\ \text{in.} \times \dfrac{2.54\ \text{cm}}{1\ \text{in.}} \times \dfrac{1\ \text{m}}{100\ \text{cm}} = 1.8\ \text{m}$

(b) $175\ \text{lb} \times \dfrac{454\ \text{g}}{1\ \text{lb}} \times \dfrac{1\ \text{kg}}{1000\ \text{g}} = 79.5\ \text{kg}$

(c) $0.500\ \text{qt} \times \dfrac{946\ \text{mL}}{1\ \text{qt}} \times \dfrac{1\ \text{L}}{1000\ \text{mL}} = 0.473\ \text{L}$

(d) $4.55 \times 10^{-4}\ \text{min} \times \dfrac{60\ \text{s}}{1\ \text{min}} \times \dfrac{10\ \text{ds}}{1\ \text{s}} = 0.273\ \text{ds}$

25. $1776\ \text{ft} \times \dfrac{12\ \text{in.}}{1\ \text{ft}} \times \dfrac{2.54\ \text{cm}}{1\ \text{in.}} \times \dfrac{1\ \text{m}}{100\ \text{cm}} = 541\ \text{m}$

27. $\dfrac{19\ \text{km}}{1\ \text{L}} \times \dfrac{0.621\ \text{mi}}{1\ \text{km}} \times \dfrac{1\ \text{L}}{0.264\ \text{gal}} = 45\ \text{mi/gal}$

Section 2.5 *The Percent Concept*

29. $65.5\ \text{g sterling silver} \times \dfrac{7.50\ \text{g copper}}{100\ \text{g sterling silver}} = 4.91\ \text{g copper}$

31. $255\ \text{mL ethanol} + 375\ \text{mL water} = 630\ \text{mL} = 6.30 \times 10^2\ \text{mL solution}$

$$\frac{255\ \text{mL}}{6.30 \times 10^2\ \text{mL}} \times 100\% = 40.5\%\ \text{ethanol in gasohol}$$

33. mass of coin = 2.554 g copper and zinc
mass of zinc = 2.490 g zinc
mass of copper = 2.554 g – 2.490 g = 0.064 g copper

Percent copper: $\dfrac{0.064\ \text{g}}{2.554\ \text{g}} \times 100\% = 2.5\%$ copper in 1990 penny

Section 2.6 *Volume by Calculation*

35. (a) $1\ \text{mL} \times \dfrac{1\ \text{cm}^3}{1\ \text{mL}} = 1\ \text{cm}^3$ (exactly)

(b) $1\ \text{in.}^3 \times \dfrac{2.54\ \text{cm}}{1\ \text{in.}} \times \dfrac{2.54\ \text{cm}}{1\ \text{in.}} \times \dfrac{2.54\ \text{cm}}{1\ \text{in.}} = 16.4\ \text{cm}^3$

37. $3.50\ \text{mm} \times \dfrac{1\ \text{m}}{1000\ \text{mm}} \times \dfrac{100\ \text{cm}}{1\ \text{m}} = 0.350\ \text{cm}$

$2.50\ \text{cm} \times 1.25\ \text{cm} \times 0.350\ \text{cm} = 1.09\ \text{cm}^3$

39. $\dfrac{(5.00\ \text{cm}^3)}{(75.0\ \text{cm})\ (35.0\ \text{cm})} = \dfrac{(5.00\ \text{cm}\ \times \cancel{\text{cm}}\ \times \cancel{\text{cm}})}{(75.0\ \cancel{\text{cm}})\ (35.0\ \cancel{\text{cm}})} = 1.90\ \times 10^{-3}\ \text{cm}$

41. $355\ \cancel{\text{in.}}^3 \times \left(\dfrac{2.54\ \cancel{\text{cm}}}{1\ \cancel{\text{in.}}}\right)^3 \times \dfrac{1\ \cancel{\text{mL}}}{1\ \cancel{\text{cm}^3}} \times \dfrac{1\ \text{L}}{1000\ \cancel{\text{mL}}} = 5.82\ \text{L}$

Section 2.7 *Volume by Displacement*

43. volume of gemstone: 7.0 mL – 4.5 mL = 2.5 mL

45. volume of hydrogen gas: 95.0 mL – 32.5 mL = 62.5 mL

Section 2.8 *The Density Concept*

47.

	Substance	Sink or Float		Substance	Sink or Float
(a)	wax	float	(b)	marble	sink

49.

	Gas	Rise or Drop		Gas	Rise or Drop
(a)	helium	rise	(b)	argon	drop

51. The density of water is 1.00 g/mL.

53. (a) $\dfrac{19.7\ \text{g}}{25.0\ \cancel{\text{mL}}} = 0.788\ \text{g/mL}$

(b) $\dfrac{10.0\ \text{g}}{14.0\ \text{mL}} = 0.714\ \text{g/mL}$

55. (a) $250\ \cancel{\text{mL}} \times \dfrac{0.69\ \text{g}}{1\ \cancel{\text{mL}}} = 170\ \text{g}$

(b) $150\ \cancel{\text{mL}} \times \dfrac{0.79\ \text{g}}{1\ \cancel{\text{mL}}} = 120\ \text{g}$

57. (a) $25.0\ \cancel{\text{g}} \times \dfrac{1\ \text{mL}}{0.714\ \cancel{\text{g}}} = 35.0\ \text{mL}$

(b) $15.0\ \cancel{\text{g}} \times \dfrac{1\ \text{mL}}{0.792\ \cancel{\text{g}}} = 18.9\ \text{mL}$

Section 2.9 *Temperature*

59.

	Temperature Scale	Freezing Point of Water
(a)	Fahrenheit	32 °F
(b)	Celsius	0 °C
(c)	Kelvin	273 K

61. (a) $(101 - 32)\,\cancel{°F} \times \frac{100\ °C}{180\ \cancel{°F}} = 38\ °C$

(b) $(-215 - 32)\,\cancel{°F} \times \frac{100\ °C}{180\ \cancel{°F}} = -137\ °C$

63. (a) $495\ °C + 273 = 768\ K$

(b) $-185\ °C + 273 = 88\ K$

65. $184\ K - 273 = -89\ °C$

$\left[-89\ \cancel{°C} \times \frac{180\ °F}{100\ \cancel{°C}}\right] + 32\ °F = -128\ °F$

Section 2.10 *The Heat Concept*

67. In Table 2.5, gold is listed as the solid that is the best conductor of heat.

69. $450\ \cancel{kcal} \times \frac{1000\ \cancel{cal}}{1\ \cancel{kcal}} \times \frac{4.184\ \cancel{J}}{1\ \cancel{cal}} \times \frac{1\ kJ}{1000\ \cancel{J}} = 1900\ kJ$

71. $25.0\ \cancel{g} \times \frac{0.108\ cal}{1\ g \times °C} \times (50.0 - 25.0)\ \cancel{°C} = 67.5\ cal$

73. $\frac{25.5\ cal}{30.0\ g \times (55.0 - 27.5)\ °C} = 0.0309\ cal / g \times °C$

75. $75.5\ \cancel{cal} \times \frac{1\ g \times \cancel{°C}}{0.125\ \cancel{cal}} \times \frac{1}{(32.0 - 20.5)\ \cancel{°C}} = 52.5\ g$ titanium

General Exercises

77. $1.5\ \cancel{TB} \times \frac{1 \times 10^{12}\ \cancel{Bytes}}{1\ \cancel{TB}} \times \frac{1\ MB}{1 \times 10^{6}\ \cancel{Bytes}} = 1.5 \times 10^{6}\ MB$

79. $1.5\ \cancel{TB} \times \frac{1 \times 10^{12}\ \cancel{Bytes}}{1\ \cancel{TB}} \times \frac{1\ \cancel{GB}}{1 \times 10^{9}\ \cancel{Bytes}} \times \frac{1\ DVD}{4.7\ \cancel{GB}} = 320\ DVDs$

81. $35\ \cancel{ms} \times \frac{1\ \cancel{s}}{1000\ \cancel{ms}} \times \frac{1 \times 10^{6}\ \mu s}{1\ \cancel{s}} = 3.5 \times 10^{4}\ \mu s$

83. $1\ \cancel{yr} \times \frac{365\ \cancel{d}}{1\ \cancel{yr}} \times \frac{24\ \cancel{h}}{1\ \cancel{d}} \times \frac{60\ \cancel{min}}{1\ \cancel{h}} \times \frac{60\ \cancel{s}}{1\ \cancel{min}} \times \frac{1.86 \times 10^{5}\ mi}{1\ \cancel{s}} = 5.87 \times 10^{12}\ mi$

85. $94.0\ \cancel{ft} \times \frac{1\ \cancel{yd}}{3\ \cancel{ft}} \times \frac{0.914\ m}{1\ \cancel{yd}} = 28.6\ m$

$50.0\ \cancel{ft} \times \frac{1\ \cancel{yd}}{3\ \cancel{ft}} \times \frac{0.914\ m}{1\ \cancel{yd}} = 15.2\ m$

Area of a basketball court: $28.6\ m \times 15.2\ m = 435\ m^2$

87. $400\ \cancel{m} \times \frac{1\ \cancel{km}}{1000\ \cancel{m}} \times \frac{1\ mile}{1.61\ \cancel{km}} = 0.248\ mile$

Note: A 400-meter race (0.248 mi) is less than a quarter mile (0.250 mi).

89. $2.50\ \cancel{kg} \times \frac{1000\ \cancel{g}}{1\ \cancel{kg}} \times \frac{1000\ \cancel{mg}}{1\ \cancel{g}} \times \frac{1\ tablet}{325\ \cancel{mg}} = 7690\ tablets$

91. Specific gravity of gasohol: $\frac{0.801\ \cancel{g/mL}}{1.00\ \cancel{g/mL}} = 0.801$

Challenge Exercises

93. $\frac{1.00\ \cancel{g}}{1\ \cancel{mL}} \times \frac{1\ lb}{454\ \cancel{g}} \times \frac{946\ \cancel{mL}}{1\ \cancel{qt}} \times \frac{4\ \cancel{qt}}{1\ gal} = 8.33\ lb/gal$

95. By definition, the mass of the reference kilogram is exactly one kilogram (1 kg).

$V = 1\ \cancel{kg} \times \frac{1000\ \cancel{g}}{1\ \cancel{kg}} \times \frac{1\ cm^3}{21.50\ \cancel{g}} = 46.51\ cm^3$

$V = \pi r^2 h = 46.51\ cm^3$

$h = \frac{46.51\ cm^3}{\pi \cdot r^2} = \frac{46.51\ cm^3}{3.14 \times (1.95\ cm)^2} = 3.90\ cm$

CHAPTER 3

Matter and Energy

Section 3.1 *Physical States of Matter*

1. The *solid state* has a fixed shape and fixed volume.

3. The *gaseous state* has a variable shape and compresses significantly.

5. The *solid state* has closely packed particles, which are held rigidly in position.

7.

	Change of State	Term
(a)	gas to liquid	condensing
(b)	liquid to solid	freezing

9.

	Change of State	Heat/Cool
(a)	gas to liquid	cooling
(b)	liquid to solid	cooling

11. The process of *deposition* (gas to solid) requires *cooling,* and the gas releases heat in the process.

Section 3.2 *Elements, Compounds, and Mixtures*

13.

	Example	Classification
(a)	sodium metal	element
(b)	chlorine gas	element
(c)	sodium chloride	compound
(d)	salt water	mixture

15.

	Example	Classification
(a)	copper wire	element
(b)	copper ore	heterogeneous mixture
(c)	copper oxide	compound
(d)	brass alloy	homogeneous mixture

Section 3.3 *Names and Symbols of the Elements*

17.

	Element	Symbol		Element	Symbol
(a)	lithium	Li	(b)	argon	Ar
(c)	magnesium	Mg	(d)	manganese	Mn
(e)	beryllium	Be	(f)	silicon	Si
(g)	mercury	Hg	(h)	titanium	Ti

19.

	Symbol	Element		Symbol	Element
(a)	P	phosphorus	(b)	S	sulfur
(c)	Kr	krypton	(d)	Fe	iron
(e)	Sb	antimony	(f)	N	nitrogen
(g)	Pt	platinum	(h)	Ag	silver

21.

	Element	Atomic Number		Element	Atomic Number
(a)	hydrogen	1	(b)	boron	5
(c)	aluminum	13	(d)	titanium	22
(e)	arsenic	33	(f)	strontium	38
(g)	tin	50	(h)	bismuth	83

Section 3.4 *Metals, Nonmetals, and Semimetals*

23.

	Property	Classification
(a)	shiny luster	metal
(b)	low melting point	nonmetal
(c)	malleable	metal
(d)	reacts with metals and nonmetals	nonmetal

25.

	Property	Classification
(a)	dull powder	nonmetal
(b)	colorless gas	nonmetal
(c)	high density	metal
(d)	reacts with metals	nonmetal

27.

	Element	Classification
(a)	H	nonmetal
(b)	He	nonmetal
(c)	Hf	metal
(d)	Ho	metal

29.

	Element	Classification
(a)	boron	semimetal
(b)	barium	metal
(c)	beryllium	metal
(d)	bismuth	metal

31.

	Element	Physical State
(a)	N	gas
(b)	Na	solid
(c)	Ne	gas
(d)	Ni	solid

33.

	Element	Physical State
(a)	hydrogen	gas
(b)	titanium	solid
(c)	fluorine	gas
(d)	bromine	liquid

Section 3.5 *Compounds and Chemical Formulas*

35.

	Chemical Formula	Chemical Composition
(a)	$C_{27}H_{46}O$	27 carbon atoms 46 hydrogen atoms 1 oxygen atom
(b)	$C_{20}H_{24}N_2O_2$	20 carbon atoms 24 hydrogen atoms 2 nitrogen atoms 2 oxygen atoms

37.

	Chemical Composition	Chemical Formula
(a)	6 carbon, 14 hydrogen, 2 nitrogen, and 2 oxygen atoms	lysine: $C_6H_{14}N_2O_2$
(b)	5 carbon, 11 hydrogen, 1 nitrogen, 2 oxygen, and 1 sulfur atom	methionine: $C_5H_{11}NO_2S$

39.

	Chemical Formula	Total Atoms
(a)	$C_3H_4OH(COOH)_3$	citric acid: 21 atoms
(b)	$C_2H_4(OH)COOH$	lactic acid: 12 atoms

41. From the law of constant composition, we can predict the mass ratio of the elements in water is constant: 1:8.

Section 3.6 *Physical and Chemical Properties*

43.

	Property		Property
(a)	physical	(b)	physical
(c)	physical	(d)	chemical

45.

	Property		Property
(a)	chemical	(b)	physical
(c)	chemical	(d)	physical

47.

	Property		Property
(a)	chemical	(b)	physical
(c)	chemical	(d)	physical

49.

	Property		Property
(a)	chemical	(b)	physical
(c)	chemical	(d)	chemical

Section 3.7 *Physical and Chemical Changes*

51.

	Change		Change
(a)	chemical	(b)	physical
(c)	physical	(d)	chemical

53.

	Change		Change
(a)	physical	(b)	chemical
(c)	physical	(d)	chemical

55.

	Change		Change
(a)	chemical	(b)	physical
(c)	physical	(d)	chemical

57.

	Change		Change
(a)	physical	(b)	chemical
(c)	chemical	(d)	physical

Section 3.8 *Conservation of Mass*

59. 1.00 g zinc + 0.49 g sulfur = 1.49 g zinc sulfide

61. 10.11 g of limestone – 8.51 g of calcium oxide = 1.60 g carbon dioxide

Section 3.9 *Potential and Kinetic Energy*

63. The particles in a *solid* have the lowest kinetic energy.

65. If the temperature increases, the kinetic energy will *increase.*

67. If a steel cylinder of helium gas is heated from 50 °C to 100 °C, there is more kinetic energy and the motion of helium atoms will *increase.*

Section 3.10 *Conservation of Energy*

69. From the law of conservation of energy, 10.5 kcal of heat energy is lost when water freezes to ice at 0 °C.

71. 697 cal + 110 cal = 807 cal released as the mercury vapor cools to a liquid from 356 °C to 25 °C.

73. From the law of conservation of energy, 68.2 kcal of heat is required to decompose 18.0 g of water into hydrogen and oxygen gases.

75.

	Energy Conversion	Forms of Energy
(a)	burning coal converts water to steam	chemical → heat
(b)	steam drives a turbine	heat → mechanical
(c)	a turbine turns a generator	mechanical → mechanical
(d)	a generator makes electricity	mechanical → electrical

77.

	Energy Conversion	Forms of Energy
(a)	hot gases moving pistons	heat → mechanical
(b)	pistons powering the crankshaft	mechanical → mechanical
(c)	crankshaft turning the generator	mechanical → mechanical
(d)	generator charging the battery	mechanical → electrical

General Exercises

79. A pre-1982 penny is an alloy, and thus an example of a *homogeneous mixture.*

81. Gasoline is a mixture of compounds, but it is a *homogeneous mixture* because the properties are constant throughout a sample.

83. The element *germanium* is below silicon in Group IV/14; thus germanium has similar chemical properties and may be substituted for silicon.

85. The three most abundant elements in Earth's crust are *oxygen, silicon,* and *aluminum.*

87.

	Element	Atomic Number
(a)	ferrum	26
(b)	cuprum	29
(c)	stannum	50
(d)	plumbum	82

89. From the conservation of energy law, we know that the energy to decompose a compound is equal to the energy released when the elements react and combine. Thus, *3250 cal* of heat is required to decompose 5.00 g of ammonia into hydrogen gas and nitrogen gas.

91. According to the conservation of mass and energy law, the products will weigh *slightly less* than the reactants. In theory, a small amount of mass is converted into heat energy; in practice, the *mass loss* is undetectable.

93. In Einstein's equation: E = energy, m = mass, and c = the speed of light.

Challenge Exercises

95. A *physical change* involves a change of physical state, but not a change of chemical formula; for example, H_2O is the formula for ice, water, and steam.

CHAPTER

4 Models of the Atom

Section 4.1 *Dalton Model of the Atom*

1. Empedocles suggested the concept of an atom in 486 B.C.

3. Dalton cited the work of (1) *Robert Boyle,* who first proposed the particle nature of gases; (2) *Antoine Lavoisier,* who established the law of conservation of mass; and (3) *Joseph Proust,* who established the law of definite composition.

5. (c) "Atoms are indivisible" proved to be *invalid* with the discovery of subatomic particles; that is, electron, proton, and neutron.

Section 4.2 *Thomson Model of the Atom*

7. Raisins in plum pudding represent *electrons* in atoms.

9. The *electron* (e^-) is the simplest negative particle.

11. The relative charge on an electron (e^-) is minus one.

Section 4.3 *Rutherford Model of the Atom*

13. Protons and neutrons are located in the atomic nucleus.

15. The approximate size of an atom is 1×10^{-8} cm.

17. The relative mass of an electron is 1/1836 to that of a proton.

Section 4.4 *Atomic Notation*

19.

	Isotope	Neutrons		Isotope	Neutrons
(a)	${}^{1}_{1}H$	$1 - 1 = 0\ n^0$	(b)	${}^{4}_{2}He$	$4 - 2 = 2\ n^0$
(c)	${}^{16}_{8}O$	$16 - 8 = 8\ n^0$	(d)	${}^{20}_{10}Ne$	$20 - 10 = 10\ n^0$

21.

	Isotope	Neutrons		Isotope	Neutrons
(a)	hydrogen-1	$1 - 1 = 0\ n^0$	(b)	carbon-13	$13 - 6 = 7\ n^0$
(c)	cobalt-59	$59 - 27 = 32\ n^0$	(d)	iodine-127	$127 - 53 = 74\ n^0$

23.

Atomic Notation	Atomic Number	Mass Number	Number of Protons	Number of Neutrons	Number of Electrons
${}^{11}_{5}\text{B}$	5	11	5	6	5
${}^{15}_{7}\text{N}$	7	15	7	8	7
${}^{40}_{20}\text{Ca}$	20	40	20	20	20
${}^{200}_{80}\text{Hg}$	80	200	80	120	80

25.

(a) 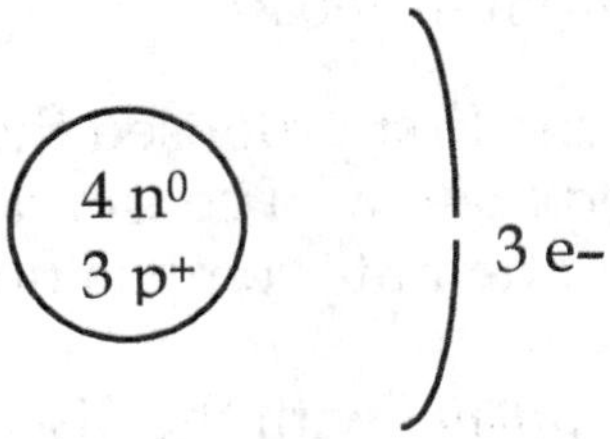

(b) 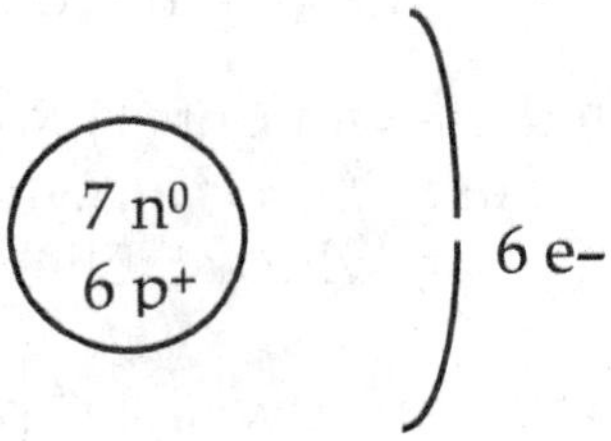

(c) 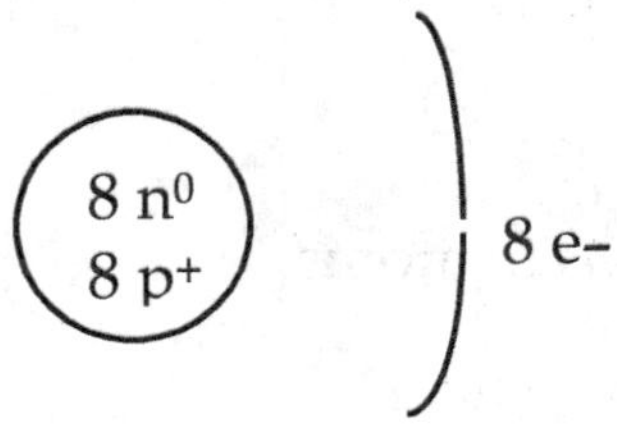

(d) 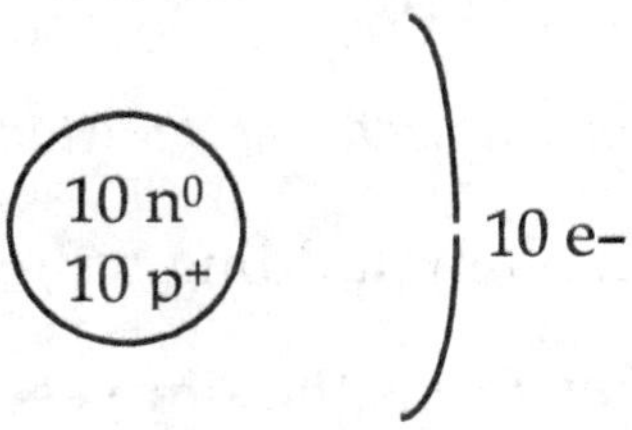

Section 4.5 *Atomic Mass*

27. The *atomic number* refers to the number of protons in the nucleus; whereas, the mass number refers to the number of protons *and* neutrons.

29. The assigned mass for the carbon-12 reference isotope is exactly 12 amu.

31. ^{23}Na has only one naturally occurring isotope with a mass of 22.99 amu.

33. ^{19}F has only one naturally occurring isotope with a mass of 19.00 amu.

35.

Li-6:	6.015 amu × 0.0742	=	0.446 amu
Li-7:	7.016 amu × 0.9258	=	6.495 amu
	Atomic Mass	=	6.941 amu

37.

Fe-54:	53.940 amu × 0.0582	=	3.14 amu
Fe-56:	55.935 amu × 0.9166	=	51.27 amu
Fe-57:	56.935 amu × 0.0219	=	1.25 amu
Fe-58:	57.933 amu × 0.0033	=	0.19 amu
	Atomic Mass	=	55.85 amu

Section 4.6 *The Wave Nature of Light*

39. *Red light* has a longer wavelength than yellow light.

41. *Yellow light* has a higher frequency than red light.

43. *Yellow light* has higher energy than red light.

45. A wavelength of *400 nm* has higher energy than 450 nm.

Section 4.7 *The Quantum Concept*

47. A *photon* represents the quantum nature of light energy.

49.

	Example	Spectrum
(a)	rainbow	continuous
(b)	line spectrum	quantized

51.

	Example	Spectrum
(a)	metric ruler	continuous
(b)	digital laser	quantized

Section 4.8 *Bohr Model of the Atom*

53. The transition from energy level 5 to 2 is most energetic because the electron drops from a higher energy level than 4 to 2, or 3 to 2.

55. The transition from energy level 4 to 1 is most energetic because the electron drops from a higher energy level than 3 to 1, or 2 to 1.

57.

Energy Level Change	Spectral Line Color
3 to 2	red line

59.

Energy Level Change	Type of Emission
4 to 1	ultraviolet energy

61. The *red line* in the emission spectrum of hydrogen has the least energy because a red photon corresponds to the electron dropping the least distance (3 to 2).

63.

	Energy Level Change	Number of Photons
(a)	3 to 1	1
(b)	3 to 2	1

65.

	Energy Level Change	Spectral Line Color
(a)	2 to 1	ultraviolet line
(b)	3 to 2	red line
(c)	4 to 3	infrared line

Section 4.9 *Energy Levels and Sublevels*

67. The lines in the emission spectrum of hydrogen suggest energy levels.

69.

	Energy Level	Number of Sublevels
(a)	1st level	1 sublevel
(b)	2nd level	2 sublevels
(c)	3rd level	3 sublevels
(d)	4th level	4 sublevels

71.

	Sublevel	Max. Electrons
(a)	$2s$	2 e–
(b)	$4p$	6 e–
(c)	$3d$	10 e–
(d)	$5f$	14 e–

73. The maximum number of electrons in the second energy level is equal to the sum of the maximum number of electrons in the $2s$ and $2p$ sublevels, that is, $2\ e^- + 6\ e^- = 8\ e^-$.

Section 4.10 *Electron Configuration*

75. The $3d$ sublevel follows the $4s$ sublevel.

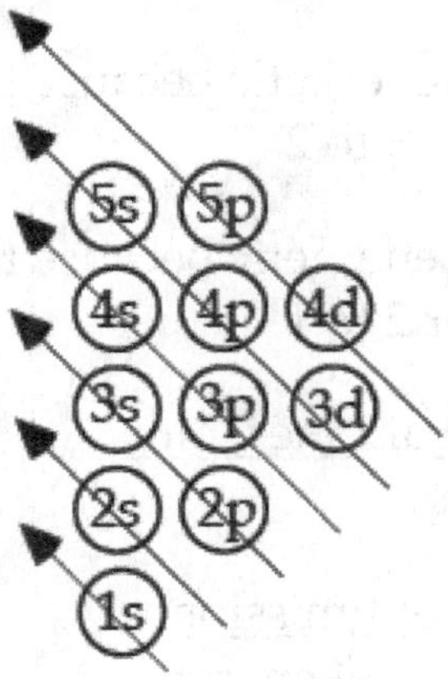

77.

	Element	Electron Configuration
(a)	He	$1s^2$
(b)	Be	$1s^2 2s^2$
(c)	Co	$1s^2 2s^2 2p^6 3s^2 3p^6 4s^2 3d^7$
(d)	Cd	$1s^2 2s^2 2p^6 3s^2 3p^6 4s^2 3d^{10} 4p^6 5s^2 4d^{10}$

79.

	Electron Configuration	Element
(a)	$1s^2 2s^1$	Li
(b)	$1s^2 2s^2 2p^6 3s^2 3p^2$	Si
(c)	$1s^2 2s^2 2p^6 3s^2 3p^6 4s^2 3d^2$	Ti
(d)	$1s^2 2s^2 2p^6 3s^2 3p^6 4s^2 3d^{10} 4p^6 5s^2$	Sr

Section 4.11 *Quantum Mechanical Model of the Atom*

81. (a)

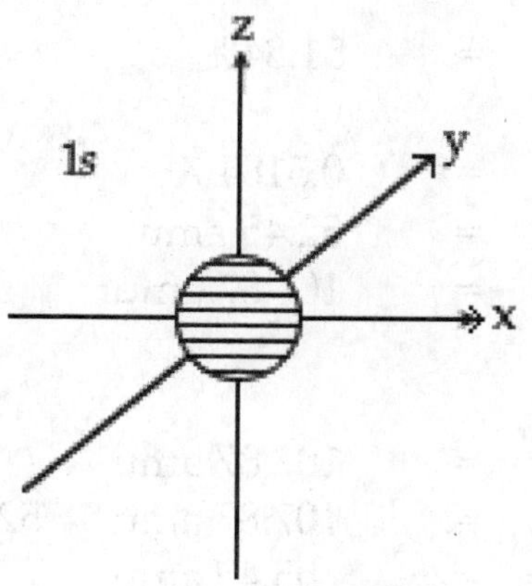

(b)

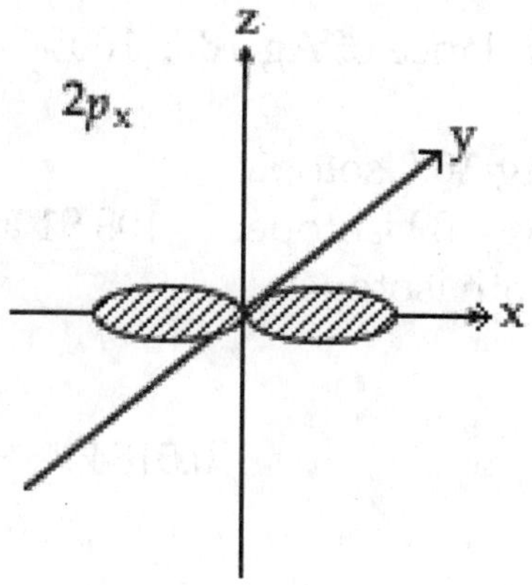

(c)

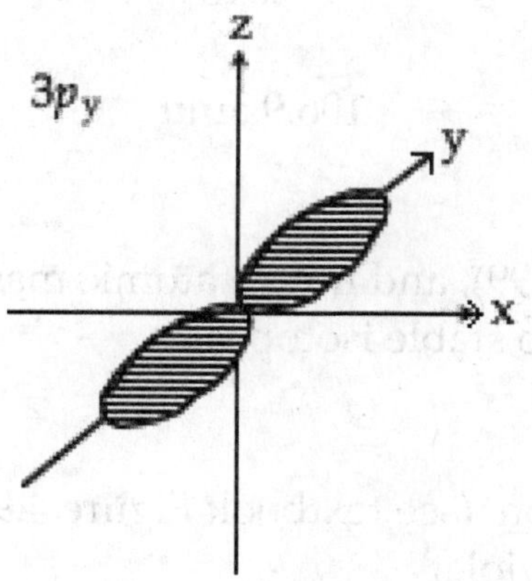

(d)

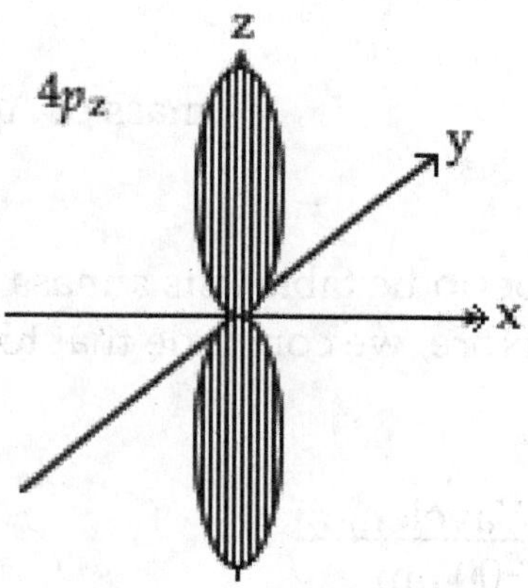

83.

	Orbitals	Higher Energy		Orbitals	Higher Energy
(a)	2*s* or 3*s*	3*s*	(b)	$2p_x$ or $3p_x$	$3p_x$
(c)	$2p_x$ or $2p_y$	both are equal	(d)	$4p_y$ or $4p_z$	both are equal

85.

	Description	Orbital
(a)	spherical orbital in the fifth energy level	5*s*
(b)	dumbbell-shaped orbital in the fourth energy level	4*p*

87.

	Orbital	Max. # of Electrons
(a)	1*s* orbital	2 e^-
(b)	2*p* orbital	2 e^-
(c)	3*d* orbital	2 e^-
(d)	4*f* orbital	2 e^-

General Exercises

89. Let X = the mass of Ag-107 isotope

Abundance of Ag-107:	100% – 48.16%	=	51.84%
Ag-107 isotope:	X × 0.5184	=	0.5184 X
Ag-109 isotope:	108.91 amu × 0.4816	=	52.45 amu
Both isotopes:	*atomic mass*	=	107.87 amu

$$0.5184\,X + 52.45\text{ amu} = 107.87\text{ amu}$$
$$0.5184\,X = 107.87\text{ amu} - 52.45\text{ amu}$$
$$0.5184\,X = 55.42\text{ amu}$$
$$X = \frac{55.42\text{ amu}}{0.5184}$$
$$\text{mass of Ag-107 isotope} = 106.9\text{ amu}$$

91. The periodic table lists a mass number for Tc (99), and not an atomic mass. Therefore, we conclude that technetium has no stable isotopes.

93.

	Wavelength	Region (see textbook Figure 4.9)
(a)	300 nm	ultraviolet
(b)	1000 nm	infrared

Challenge Exercises

95. If the average mass of silver is approximately 108 amu and one isotope is Ag-109, given an equal abundance the other isotope must be Ag-107.

97. An atom is more stable with a completely filled *d* sublevel than a partially filled sublevel. If one of the silver 5*s* electrons drops into the 4*d* sublevel, the 4*d* sublevel is completely filled, and thus the silver atom is more stable.

CHAPTER

5

The Periodic Table

Section 5.1 *Classification of Elements*

1. Neon was not discovered until 1898 (see textbook Figure 3.7).

3. Mendelevium was not synthesized until 1955 (see textbook Figure 3.7).

Section 5.2 *The Periodic Law Concept*

5. Before Moseley's discovery in 1913, the periodic law stated that physical and chemical properties tend to repeat periodically when elements are arranged according to *increasing atomic mass.*

7. Note that *Co* precedes *Ni* in the periodic table, even though the atomic mass of Co (58.93) is greater than Ni (58.69).

Section 5.3 *Groups and Periods of Elements*

9. Vertical columns in the periodic table are called *groups* or *families.*

11. The term for the elements that belong to Groups IA through VIIIA is the *representative elements.*

13. The elements in the series that follow element 57 are called the *lanthanides.*

15. The term for the elements that include Ce–Lu is the *lanthanides.*

17. The elements on the left side of the periodic table are *metals.*

19. The elements Sc and Y are *rare earth elements,* but not *lanthanides.*

21.

	Family	Group Number
(a)	alkali metals	IA/1
(b)	alkaline earth metals	IIA/2
(c)	halogens	VIIA/17
(d)	noble gases	VIIIA/18

23.

	American	IUPAC		American	IUPAC
(a)	Group IA	1	(b)	Group IIIA	13
(c)	Group VA	15	(d)	Group VIIA	17
(e)	Group IB	11	(f)	Group IIIB	3
(g)	Group VB	5	(h)	Group VIIB	7

25.

	Element		Element
(a)	B	(b)	Ge
(c)	Cl	(d)	At

27.

	Element		Element
(a)	Na	(b)	Mg
(c)	Lu	(d)	Pm

Section 5.4 *Periodic Trends*

29. Proceeding up a group of elements in the periodic table, the metallic character generally *decreases.*

31. Proceeding up a group of elements in the periodic table, the atomic radius generally *decreases.*

33.

	Metallic Character		Metallic Character
(a)	**K** > Ca	(b)	**Mg** > Al
(c)	**Fe** > Cu	(d)	**S** > Ar

(Note: The element with the greater metallic character is in **bold**.)

35.

	Atomic Radius		Atomic Radius
(a)	Li < **Na**	(b)	N < **P**
(c)	Mg < **Ca**	(d)	Ar < **Kr**

(Note: The element with the larger atomic radius is in **bold**.)

Section 5.5 *Properties of Elements*

37. Predicted atomic radius for Fr: 248 pm + (266 – 248) pm = 284 pm

    Predicted density for Fr: 1.93 g/mL + (1.93 – 1.53) g/mL = 2.33 g/mL

    Predicted melting point for Fr: 28.4 °C - (39.3 – 28.4) °C = 17.5 °C

39. Predicted atomic radius for At: 133 pm + (133 – 115) pm = 151 pm

    Predicted density for At: 4.93 g/mL + (4.93 – 3.10) g/mL = 6.76 g/mL

    Predicted boiling point for At: 184 °C + (184 – 59) °C = 309 °C

41.

	Compound	Formula		Compound	Formula
(a)	potassium fluoride	KF	(b)	calcium fluoride	CaF_2
(c)	boron bromide	BBr_3	(d)	germanium iodide	GeI_4

43.

	Compound	Formula		Compound	Formula
(a)	calcium fluoride	CaF_2	(b)	calcium chloride	$CaCl_2$
(c)	calcium bromide	$CaBr_2$	(d)	calcium iodide	CaI_2

45.

	Compound	Formula		Compound	Formula
(a)	arsenic oxide	As_2O_3	(b)	antimony oxide	Sb_2O_3
	and,	As_2O_5		and,	Sb_2O_5

Section 5.6 *Blocks of Elements*

47. The elements in Groups IA/1 and IIA/2 are filling *s* sublevels.

49. The transition elements are filling *d* sublevels.

51. The elements in the lanthanide series are filling a 4*f* sublevel.

53.

	Element	Highest Sublevel		Element	Highest Sublevel
(a)	H	1*s*	(b)	Na	3*s*
(c)	Sm	4*f*	(d)	Br	4*p*
(e)	Sr	5*s*	(f)	C	2*p*
(g)	Sn	5*p*	(h)	Cs	6*s*

55.

	Element	Electron Configuration	Core Notation
(a)	Li	$1s^2\,2s^1$	[He] $2s^1$
(b)	F	$1s^2\,2s^2\,2p^5$	[He] $2s^2\,2p^5$
(c)	Mg	$1s^2\,2s^2\,2p^6\,3s^2$	[Ne] $3s^2$
(d)	P	$1s^2\,2s^2\,2p^6\,3s^2\,3p^3$	[Ne] $3s^2\,3p^3$
(e)	Ca	$1s^2\,2s^2\,2p^6\,3s^2\,3p^6\,4s^2$	[Ar] $4s^2$
(f)	Mn	$1s^2\,2s^2\,2p^6\,3s^2\,3p^6\,4s^2\,3d^5$	[Ar] $4s^2\,3d^5$
(g)	Ga	$1s^2\,2s^2\,2p^6\,3s^2\,3p^6\,4s^2\,3d^{10}\,4p^1$	[Ar] $4s^2\,3d^{10}\,4p^1$
(h)	Rb	$1s^2\,2s^2\,2p^6\,3s^2\,3p^6\,4s^2\,3d^{10}\,4p^6\,5s^1$	[Kr] $5s^1$

Section 5.7 *Valence Electrons*

57.

	Group	Valence Electrons		Group	Valence Electrons
(a)	IA/1	1	(b)	IIIA/13	3
(c)	VA/15	5	(d)	VIIA/17	7

59.

	Element	Valence Electrons		Element	Valence Electrons
(a)	H	1	(b)	B	3
(c)	N	5	(d)	F	7
(e)	Ca	2	(f)	Si	4
(g)	O	6	(h)	Ar	8

Section 5.8 *Electron Dot Formulas*

61.

	Element	Electron Dot Formula		Element	Electron Dot Formula
(a)	H	H·	(b)	B	·B· (one dot below)
(c)	N	·N: (one dot above, one below)	(d)	F	:F: (one dot above, two below)
(e)	Ca	Ca· (one dot below)	(f)	Si	·Si· (one dot above, one below)
(g)	O	·O: (one dot above, two below)	(h)	Ar	:Ar: (two dots above, two below)

Section 5.9 *Ionization Energy*

63. Proceeding up a group in the periodic table, the ionization energy *increases.*

65. The Group IA/1 elements have the *lowest* ionization energy.

67.

	Ionization Energy		Ionization Energy
(a)	**Mg** > Ca	(b)	**S** > Se
(c)	**Sn** > Pb	(d)	**N** > P

(Note: The element with the higher ionization energy is in **bold**.)

69.

	Ionization Energy		Ionization Energy
(a)	Rb > **Cs**	(b)	He > **Ar**
(c)	B > **Al**	(d)	F > **I**

(Note: The element with the lower ionization energy is in **bold**.)

Section 5.10 *Ionic Charges*

71.

	Group	Ionic Charge		Group	Ionic Charge
(a)	IA/1	1+	(b)	IIA/2	2+
(c)	IIIA/13	3+	(d)	IVA/14	4+

73.

	Ion	Ionic Charge		Ion	Ionic Charge
(a)	Be ion	2+	(b)	Sn ion	4+
(c)	P ion	3–	(d)	S ion	2–

75.

	Ion	Isoelectronic		Ion	Isoelectronic
(a)	K^+	Ar	(b)	$\mathbf{Sr^{2+}}$	Kr
(c)	Cl^-	Ar	(d)	$\mathbf{Se^{2-}}$	Kr

77.

	Ion	Electron Configuration – Core Notation
(a)	Mg^{2+}	[Ne]
(b)	K^+	[Ar]
(c)	Fe^{2+}	[Ar] $3d^6$
(d)	Zr^{2+}	[Kr] $4d^2$

79.

	Ion	Electron Configuration – Core Notation
(a)	F^-	[He] $2s^2\ 2p^6$, or [Ne]
(b)	S^{2-}	[Ne] $3s^2\ 3p^6$, or [Ar]
(c)	N^{3-}	[He] $2s^2\ 2p^6$, or [Ne]
(d)	I^-	[Kr] $5s^2\ 4d^{10}\ 5p^6$, or [Xe]

General Exercises

81.

	European	American		European	American
(a)	Group IA	Group IA	(b)	Group IB	Group IB
(c)	Group IIIA	Group IIIB	(d)	Group IIIB	Group IIIA

83. Predicted atomic radius for Mo: $\dfrac{125 \text{ pm} + 137 \text{ pm}}{2} = 131 \text{ pm}$

Predicted density for W: 10.28 g/mL + (10.28 – 7.14) g/mL = 13.42 g/mL

Predicted melting point for Cr: 2617 °C – (3410 – 2617) °C = 1824 °C

85.

	Element	Electron Configuration
(a)	Sr	[Kr] $5s^2$
(b)	Ru	[Kr] $5s^2\ 4d^6$
(c)	Sb	[Kr] $5s^2\ 4d^{10}\ 5p^3$
(d)	Cs	[Xe] $6s^1$

87. Although H and the Group IA/1 metals each have one valence electron, in H the electron is closer to its nucleus. Because the negatively-charged electron is closer to the positively-charged nucleus, it requires more energy to remove the electron from hydrogen than a Group IA/1 metal.

Challenge Exercises

89. Mendeleev predicted the element scandium (Sc), which was not discovered until 1879.

91. When an alkali metal atom loses one electron, it gains a noble gas configuration. An alkaline earth metal atom must lose two electrons in order to achieve an extremely stable, noble gas, electron configuration.

CHAPTER 6

Language of Chemistry

Section 6.1 *Classification of Compounds*

1.

	Compound	Classification		Compound	Classification
(a)	H_2O	binary molecular	(b)	Fe_2O_3	binary ionic
(c)	$CaCO_3$	ternary ionic	(d)	$HCl(aq)$	binary acid

3.

	Ion	Classification		Ion	Classification
(a)	NH_4^+	polyatomic cation	(b)	Al^{3+}	monoatomic cation
(c)	Cl^-	monoatomic anion	(d)	ClO_3^-	polyatomic anion

Section 6.2 *Monoatomic Ions*

5.

	Monoatomic Cation	Stock System Name
(a)	Ca^{2+}	calcium ion
(b)	Cd^{2+}	cadmium ion
(c)	Co^{2+}	cobalt(II) ion
(d)	Cu^{2+}	copper(II) ion

7.

	Monoatomic Anion	Systematic Name
(a)	F^-	fluoride ion
(b)	Cl^-	chloride ion
(c)	Br^-	bromide ion
(d)	I^-	iodide ion

9.

	Monoatomic Cation	Chemical Formula
(a)	potassium ion	K^+
(b)	barium ion	Ba^{2+}
(c)	iron(II) ion	Fe^{2+}
(d)	tin(II) ion	Sn^{2+}

Section 6.3 *Polyatomic Ions*

11.

	Oxyanion	Systematic Name
(a)	NO_3^-	nitrate ion
(b)	NO_2^-	nitrite ion
(c)	ClO_3^-	chlorate ion
(d)	ClO_2^-	chlorite ion

13.

	Polyatomic Ion	Chemical Formula
(a)	chromate ion	CrO_4^{2-}
(b)	dichromate ion	$Cr_2O_7^{2-}$
(c)	hydroxide ion	OH^-
(d)	cyanide ion	CN^-

Section 6.4 *Writing Chemical Formulas*

15.

	Constituent Ions	Chemical Formula
(a)	Al^{3+} + Br^-	$AlBr_3$
(b)	Cd^{2+} + S^{2-}	CdS
(c)	Mn^{2+} + P^{3-}	Mn_3P_2
(d)	Pb^{4+} + F^-	PbF_4

17.

	Constituent Ions	Chemical Formula
(a)	K^+ + NO_3^-	KNO_3
(b)	NH_4^+ + $Cr_2O_7^{2-}$	$(NH_4)_2Cr_2O_7$
(c)	Cu^+ + CO_3^{2-}	Cu_2CO_3
(d)	Mn^{2+} + SO_3^{2-}	$MnSO_3$

19.

	Constituent Ions	Chemical Formula
(a)	Sr^{2+} + NO_2^-	$Sr(NO_2)_2$
(b)	Zn^{2+} + MnO_4^-	$Zn(MnO_4)_2$
(c)	Ca^{2+} + CrO_4^{2-}	$CaCrO_4$
(d)	Ni^{2+} + ClO_4^-	$Ni(ClO_4)_2$

Section 6.5 *Binary Ionic Compounds*

21.

	Binary Ionic Compound	Systematic Name
(a)	MgO	magnesium oxide
(b)	ZnO	zinc oxide
(c)	CdO	cadmium oxide
(d)	BaO	barium oxide

23.

	Binary Ionic Compound	Systematic Name
(a)	CoO	cobalt(II) oxide
(b)	FeO	iron(II) oxide
(c)	HgO	mercury(II) oxide
(d)	SnO	tin(II) oxide

25\.

	Binary Ionic Compound	Chemical Formula
(a)	copper(I) oxide	Cu_2O
(b)	iron(II) nitride	Fe_3N_2
(c)	mercury(II) chloride	$HgCl_2$
(d)	lead(IV) sulfide	PbS_2

27\.

	Binary Ionic Compound	Chemical Formula
(a)	rubidium chloride	$RbCl$
(b)	sodium bromide	$NaBr$

29\.

	Binary Ionic Compound	Chemical Formula
(a)	gallium nitride	GaN
(b)	aluminum arsenide	$AlAs$

Section 6.6 *Ternary Ionic Compounds*

31\.

	Ternary Ionic Compound	Systematic Name
(a)	$LiMnO_4$	lithium permanganate
(b)	$Sr(ClO_2)_2$	strontium chlorite
(c)	$BaCO_3$	barium carbonate
(d)	$(NH_4)_2Cr_2O_7$	ammonium dichromate

33\.

	Ternary Ionic Compound	Stock System Name
(a)	$CuSO_4$	copper(II) sulfate
(b)	$FeCrO_4$	iron(II) chromate
(c)	$Hg(NO_2)_2$	mercury(II) nitrite
(d)	$Pb(C_2H_3O_2)_2$	lead(II) acetate

35\.

	Ternary Ionic Compound	Chemical Formula
(a)	manganese(II) acetate	$Mn(C_2H_3O_2)_2$
(b)	lead(IV) chlorite	$Pb(ClO_2)_4$
(c)	tin(II) phosphate	$Sn_3(PO_4)_2$
(d)	iron(III) hypochlorite	$Fe(ClO)_3$

37\.

	Ternary Ionic Compound	Chemical Formula
(a)	francium sulfate	Fr_2SO_4
(b)	sodium sulfite	Na_2SO_3

39\.

	Ternary Ionic Compound	Chemical Formula
(a)	radium chlorate	$Ra(ClO_3)_2$
(b)	barium bromate	$Ba(BrO_3)_2$

Section 6.7 *Latin System of Nomenclature*

41.

	Monoatomic Cation	Latin System Name
(a)	Co^{2+}	cobaltous ion
(b)	Co^{3+}	cobaltic ion
(c)	Sn^{2+}	stannous ion
(d)	Sn^{4+}	stannic ion

43.

	Monoatomic Cation	Chemical Formula
(a)	cuprous ion	Cu^{+}
(b)	cupric ion	Cu^{2+}
(c)	plumbous ion	Pb^{2+}
(d)	plumbic ion	Pb^{4+}

45.

	Ionic Compound	Latin System Name
(a)	CuS	cupric sulfide
(b)	FeO	ferrous oxide
(c)	$Hg(NO_2)_2$	mercuric nitrite
(d)	$Pb(C_2H_3O_2)_2$	plumbous acetate

47.

	Ionic Compound	Chemical Formula
(a)	cupric chloride	$CuCl_2$
(b)	ferrous sulfide	FeS
(c)	cobaltic acetate	$Co(C_2H_3O_2)_3$
(d)	stannous chromate	$SnCrO_4$

Section 6.7 *Binary Molecular Compounds*

49.

	Binary Molecular Compound	Chemical Formula
(a)	dinitrogen pentaoxide	N_2O_5
(b)	carbon tetrachloride	CCl_4
(c)	iodine monobromide	IBr
(d)	dihydrogen sulfide	H_2S

51.

	Binary Molecular Compound	Systematic Name
(a)	BrF	bromine monofluoride
(b)	CF_4	carbon tetrafluoride
(c)	I_2O_4	diiodine tetraoxide
(d)	Cl_2O_3	dichlorine trioxide

Section 6.8 *Binary Acids*

53.

	Binary Acid	Chemical Formula
(a)	hydrosulfuric acid	$H_2S(aq)$
(b)	hydrobromic acid	$HBr(aq)$

Section 6.9 *Ternary Oxyacids*

55.

	Ternary Oxyacid	Chemical Formula
(a)	acetic acid	$HC_2H_3O_2(aq)$
(b)	nitric acid	$HNO_3(aq)$

57.

	Ternary Oxyacid	Systematic Name
(a)	$HClO_2(aq)$	chlorous acid
(b)	$H_3PO_4(aq)$	phosphoric acid

General Exercises

59.

	Substance	Ionic Charge
(a)	iron metal atoms	0
(b)	ferrous ions	$2+$
(c)	iron(III) ions	$3+$
(d)	iron compounds	0

(Note: The total ionic charge on a compound is zero.)

61.

	Polyatomic Anion	Valence Electrons	
(a)	$IO_4^{?-}$	$7 + 4(6) = 31\ e^-$	(thus, 1–)
(b)	$SiO_3^{?-}$	$4 + 3(6) = 22\ e^-$	(thus, 2–)

Note: The polyatomic anion with a charge of 2– is SiO_3^{2-} because the total of the valence electrons is an even number (24).

63.

IONS	chloride ion	sulfide ion	phosphide ion
copper(I) ion	$CuCl$ copper(I) chloride	Cu_2S copper(I) sulfide	Cu_3P copper(I) phosphide
cobalt(III) ion	$CoCl_3$ cobalt(III) chloride	Co_2S_3 cobalt(III) sulfide	CoP cobalt(III) phosphide
lead(IV) ion	$PbCl_4$ lead(IV) chloride	PbS_2 lead(IV) sulfide	Pb_3P_4 lead(IV) phosphide

65.

IONS	hydroxide ion	sulfate ion	phosphate ion
mercurous ion	$Hg_2(OH)_2$ mercurous hydroxide	Hg_2SO_4 mercurous sulfate	$(Hg_2)_3(PO_4)_2$ mercurous phosphate
ferrous ion	$Fe(OH)_2$ ferrous hydroxide	$FeSO_4$ ferrous sulfate	$Fe_3(PO_4)_2$ ferrous phosphate
stannous ion	$Sn(OH)_2$ stannous hydroxide	$SnSO_4$ stannous sulfate	$Sn_3(PO_4)_2$ stannous phosphate

67.

	Compound	Suffix Ending		Binary Acid	Suffix Ending
(a)	Na_2S	-ide	(b)	$H_2S(aq)$	-ic acid

69.

	Compound	Suffix Ending		Ternary Oxyacid	Suffix Ending
(a)	Na_2SO_3	-ite	(b)	$H_2SO_3(aq)$	-ous acid

71.

	Compound	Suffix Ending		Ternary Oxyacid	Suffix Ending
(a)	Na_2SO_4	-ate	(b)	$H_2SO_4(aq)$	-ic acid

73.

	Chemical Name	Chemical Formula
(a)	dihydrogen oxide (water)	H_2O
(b)	sodium hypochlorite (bleach)	$NaClO$
(c)	sodium hydroxide (caustic soda)	$NaOH$
(d)	sodium bicarbonate (baking soda)	$NaHCO_3$

75.

	Binary Compound	Systematic Name
(a)	BF_3	boron trifluoride
(b)	$SiCl_4$	silicon tetrachloride
(c)	As_2O_5	diarsenic pentaoxide
(d)	Sb_2O_3	diantimony trioxide

77.

Given Formula	Predicted Formula
ZrO_2	TiO_2

Challenge Exercises

79.

Given Formula	Predicted Formula
$LaCl_3$	$AcCl_3$

CHAPTER 7

Chemical Reactions

Section 7.1 *Evidence of Chemical Reactions*

1. (a) The violet flame is evidence of a *chemical reaction.*
   (b) The gas bubbles are evidence of a *chemical reaction.*

3. (a) The dark red solid is evidence of a *chemical reaction.*
   (b) The gray film is evidence of a *physical change.*

5. (a) The gas bubbles is evidence of a *chemical reaction.*
   (b) The insoluble solid is evidence of a *chemical reaction.*

Section 7.2 *Writing Chemical Equations*

7. $Mg(s) + Br_2(l) \rightarrow MgBr_2(s)$

9. $Zn(HCO_3)_2(s) \rightarrow ZnCO_3(s) + H_2O(g) + CO_2(g)$

11. $Cd(s) + Co(NO_3)_2(aq) \rightarrow Cd(NO_3)_2(aq) + Co(s)$

13. $LiI(aq) + AgNO_3(aq) \rightarrow AgI(s) + LiNO_3(aq)$

15. $HC_2H_3O_2(aq) + NH_4OH(aq) \rightarrow NH_4C_2H_3O_2(aq) + HOH(l)$

Section 7.3 *Balancing Chemical Equations*

17. (a) $3\ H_2(g) + N_2(g) \rightarrow 2\ NH_3(g)$
    (b) $Al_2(CO_3)_3(s) \rightarrow Al_2O_3(s) + 3\ CO_2(g)$
    (c) $Sr(s) + 2\ H_2O(l) \rightarrow Sr(OH)_2(aq) + H_2(g)$
    (d) $K_2SO_4(aq) + Ba(OH)_2(aq) \rightarrow BaSO_4(s) + 2\ KOH(aq)$
    (e) $2\ H_3PO_4(aq) + 3\ Mn(OH)_2(s) \rightarrow Mn_3(PO_4)_2(s) + 6\ HOH(l)$

19. (a) $2\,Pb(s) + O_2(g) \rightarrow 2\,PbO(s)$
(b) $2\,LiNO_3(s) \rightarrow 2\,LiNO_2(s) + O_2(g)$
(c) $Mg(s) + 2\,HC_2H_3O_2(aq) \rightarrow Mg(C_2H_3O_2)_2(aq) + H_2(g)$
(d) $Hg_2(NO_3)_2(aq) + 2\,NaBr(aq) \rightarrow Hg_2Br_2(s) + 2\,NaNO_3(aq)$
(e) $H_2CO_3(aq) + 2\,NH_4OH(aq) \rightarrow (NH_4)_2CO_3(aq) + 2\,HOH(l)$

Section 7.4 *Classifying Chemical Reactions*

21. Refer to the chemical reactions in Exercise 17.
(a) combination reaction
(b) decomposition reaction
(c) single-replacement reaction
(d) double-replacement reaction
(e) neutralization reaction

23. Refer to the chemical reactions in Exercise 19.
(a) combination reaction
(b) decomposition reaction
(c) single-replacement reaction
(d) double-replacement reaction
(e) neutralization reaction

Section 7.5 *Combination Reactions*

General Form: $A + Z \rightarrow AZ$

25. metal + oxygen gas → metal oxide
(a) $2\,Ni(s) + O_2(g) \rightarrow 2\,NiO(s)$
(b) $4\,Fe(s) + 3\,O_2(g) \rightarrow 2\,Fe_2O_3(s)$

27. nonmetal + oxygen gas → nonmetal oxide
(a) $2\,C(s) + O_2(g) \rightarrow 2\,CO(g)$
(b) $4\,P(s) + 5\,O_2(g) \rightarrow 2\,P_2O_5(s)$

29. metal + nonmetal → ionic compound
(a) $2\,Cu(s) + Cl_2(g) \rightarrow 2\,CuCl(s)$
(b) $Co(s) + S(s) \rightarrow CoS(s)$

31. (a) $4\,Cr(s) + 3\,O_2(g) \rightarrow 2\,Cr_2O_3(s)$
(b) $2\,Cr(s) + N_2(g) \rightarrow 2\,CrN(s)$

33. (a) $4\,Li + O_2 \rightarrow 2\,Li_2O$
(b) $2\,Ca + O_2 \rightarrow 2\,CaO$

35. (a) $2\,Na + I_2 \rightarrow 2\,NaI$
(b) $3\,Ba + N_2 \rightarrow Ba_3N_2$

Section 7.6 *Decomposition Reactions*

General Form: AX $\rightarrow$ A + X

37. metal hydrogen carbonate $\rightarrow$ metal carbonate + water + carbon dioxide
    (a) $2\ KHCO_3(s) \rightarrow K_2CO_3(s) + H_2O(g) + CO_2(g)$
    (b) $Sr(HCO_3)_2(s) \rightarrow SrCO_3(s) + H_2O(g) + CO_2(g)$

39. metal carbonate $\rightarrow$ metal oxide + carbon dioxide
    (a) $Co_2(CO_3)_3(s) \rightarrow Co_2O_3(s) + 3\ CO_2(g)$
    (b) $Sn(CO_3)_2(s) \rightarrow SnO_2(s) + 2\ CO_2(g)$

41. metal hydrogen carbonate $\rightarrow$ metal carbonate + water + carbon dioxide
    (a) $2\ AgHCO_3(s) \rightarrow Ag_2CO_3(s) + H_2O(g) + CO_2(g)$
    (b) $Zn(HCO_3)_2(s) \rightarrow ZnCO_3(s) + H_2O(g) + CO_2(g)$

43. oxygen-containing compounds $\rightarrow$ oxygen gas
    (a) $Ca(NO_3)_2(s) \rightarrow Ca(NO_2)_2(s) + O_2(g)$
    (b) $2\ Ag_2SO_4(s) \rightarrow 2\ Ag_2SO_3(s) + O_2(g)$

45. oxygen-containing compounds $\rightarrow$ oxygen gas
    (a) $2\ NaClO_3(s) \rightarrow 2\ NaCl(s) + 3\ O_2(g)$
    (b) $Ca(NO_3)_2(s) \rightarrow Ca(NO_2)_2(s) + O_2(g)$

Section 7.7 *The Activity Series Concept*

47.

	Element	Solution	Observation
(a)	Ag	$Cu(NO_3)_2(aq)$	no reaction: Cu > Ag
(b)	Sn	$Cu(NO_3)_2(aq)$	reaction: Sn > Cu
(c)	Co	$Cu(NO_3)_2(aq)$	reaction: Co > Cu
(d)	Mn	$Cu(NO_3)_2(aq)$	reaction: Mn > Cu

49.

	Element	Solution	Observation
(a)	Ag	$H_2SO_4(aq)$	no reaction: (H) > Ag
(b)	Sn	$H_2SO_4(aq)$	reaction: Sn > (H)
(c)	Co	$H_2SO_4(aq)$	reaction: Co > (H)
(d)	Mn	$H_2SO_4(aq)$	reaction: Mn > (H)

51.

	Element	Solution	Observation
(a)	Li	$H_2O(l)$	reaction: Li is an active metal
(b)	Mg	$H_2O(l)$	no reaction: Mg is not an active metal
(c)	Ca	$H_2O(l)$	reaction: Ca is an active metal
(d)	Al	$H_2O(l)$	no reaction: Al is not an active metal

Section 7.8 *Single-Replacement Reactions*

General Form: $\mathbf{A + BZ \rightarrow AZ + B}$
or: $A + BZ \rightarrow NR$

53. metal$_1$ + aqueous solution$_1$ $\rightarrow$ metal$_2$ + aqueous solution$_2$
(a) $2\ Al(s) + 3\ Cu(NO_3)_2(aq) \rightarrow 3\ Cu(s) + 2\ Al(NO_3)_3(aq)$
(b) $Cu(s) + Al(NO_3)_3(aq) \rightarrow NR$

55. metal$_1$ + aqueous solution$_1$ $\rightarrow$ metal$_2$ + aqueous solution$_2$
(a) $Ni(s) + Pb(C_2H_3O_2)_2(aq) \rightarrow Pb(s) + Ni(C_2H_3O_2)_2(aq)$
(b) $Pb(s) + Ni(C_2H_3O_2)_2(aq) \rightarrow NR$

57. metal + aqueous acid $\rightarrow$ aqueous solution + hydrogen gas
(a) $Mg(s) + 2\ HNO_3(aq) \rightarrow Mg(NO_3)_2(aq) + H_2(g)$
(b) $Mn(s) + 2\ HCl(aq) \rightarrow MnCl_2(s) + H_2(g)$

59. metal + water $\rightarrow$ metal hydroxide + hydrogen gas
(a) $2\ Li(s) + 2\ H_2O(l) \rightarrow 2\ LiOH(aq) + H_2(g)$
(b) $Ba(s) + 2\ H_2O(l) \rightarrow Ba(OH)_2(aq) + H_2(g)$

61. metal$_1$ + aqueous solution$_1$ $\rightarrow$ metal$_2$ + aqueous solution$_2$
(a) $Zn(s) + Pb(NO_3)_2(aq) \rightarrow Zn(NO_3)_2(aq) + Pb(s)$
(b) $Cd(s) + Fe(NO_3)_2(aq) \rightarrow NR$

63. metal + aqueous acid $\rightarrow$ aqueous solution + hydrogen gas
(a) $Zn(s) + 2\ HNO_3(aq) \rightarrow Zn(NO_3)_2(aq) + H_2(g)$
(b) $Cd(s) + 2\ HNO_3(aq) \rightarrow Cd(NO_3)_2(aq) + H_2(g)$

65. metal + water $\rightarrow$ metal hydroxide + hydrogen gas
(a) $Zn(s) + H_2O(l) \rightarrow NR$
(b) $Ba(s) + 2\ H_2O(l) \rightarrow Ba(OH)_2(aq) + H_2(g)$

Section 7.9 *Solubility Rules*

67. (a) $(NH_4)_2S$ soluble; (b) $ZnCO_3$ insoluble
(c) $SnCrO_4$ insoluble; (d) $Pb(C_2H_3O_2)_2$ soluble

69. (a) Hg_2Cl_2 insoluble; (b) $HgCl_2$ soluble
(c) $AgBr$ insoluble; (d) PbI_2 insoluble

Section 7.10 *Double-Replacement Reactions*

General Form: $\mathbf{AX + BZ \rightarrow AZ + BX}$
or: $AX + BZ \rightarrow \textit{NR}$

71. aqueous solution$_1$ + aqueous solution$_2$ → precipitate + aqueous solution$_3$
(a) $SnCl_2(aq) + K_2S(aq) \rightarrow SnS(s) + 2\,KCl(aq)$
(b) $Co(NO_3)_2(aq) + Na_2CrO_4(aq) \rightarrow CoCrO_4(s) + 2\,NaNO_3(aq)$

73. aqueous solution$_1$ + aqueous solution$_2$ → precipitate + aqueous solution$_3$
(a) $MgSO_4(aq) + BaCl_2(aq) \rightarrow BaSO_4(s) + MgCl_2(aq)$
(b) $FeSO_4(aq) + Ca(OH)_2(aq) \rightarrow Fe(OH)_2(s) + CaSO_4(aq)$

Section 7.11 *Neutralization Reactions*

General Form: $\mathbf{HX + BOH \rightarrow BX + HOH}$

75. aqueous acid + aqueous base → aqueous salt + water
(a) $H_2CO_3(aq) + 2\,LiOH(aq) \rightarrow Li_2CO_3(aq) + 2\,HOH(l)$
(b) $2\,HNO_3(aq) + Ca(OH)_2(aq) \rightarrow Ca(NO_3)_2(aq) + 2\,HOH(l)$

77. aqueous acid + aqueous base → aqueous salt + water
(a) $2\,HNO_3(aq) + Ca(OH)_2(aq) \rightarrow Ca(NO_3)_2(aq) + 2\,HOH(l)$
(b) $2\,HNO_2(aq) + Ca(OH)_2(aq) \rightarrow Ca(NO_2)_2(aq) + 2\,HOH(l)$

General Exercises

79. Not necessarily. The sum of the coefficients may or may not be the same.

81. (a) $3\,Fe(s) + 4\,H_2O(g) \rightarrow Fe_3O_4(s) + 4\,H_2(g)$
(b) $4\,FeS(s) + 7\,O_2(g) \rightarrow 2\,Fe_2O_3(s) + 4\,SO_2(g)$

83. (a) $F_2(g) + 2\,NaBr(aq) \rightarrow Br_2(l) + 2\,NaF(aq)$
(b) $Sb_2S_3(s) + 6\,HCl(aq) \rightarrow 2\,SbCl_3(aq) + 3\,H_2S(g)$

85. (a) $CH_4(g) + 2\,O_2(g) \rightarrow CO_2(g) + 2\,H_2O(g)$
(b) $C_3H_8(g) + 5\,O_2(g) \rightarrow 3\,CO_2(g) + 4\,H_2O(g)$

Challenge Exercises

87. **Industrial Preparation of Iron:**

$Fe_2O_3(l) + 3\,CO(g) \rightarrow 2\,Fe(l) + 3\,CO_2(g)$

CHAPTER

8

The Mole Concept

Section 8.1 *Avogadro's Number*

1.

	Element	Average Mass		Element	Average Mass
(a)	Na	22.99 amu	(b)	Sr	87.62 amu
(c)	Si	28.09 amu	(d)	Se	78.97 amu

3.

	Element	Mass		Element	Mass
(a)	Na	22.99 g	(b)	Sr	87.62 g
(c)	Si	28.09 g	(d)	Se	78.97 g

Section 8.2 *Mole Calculations I*

5. (a) 1 mol C = 6.02×10^{23} atoms C
   (b) 1 mol CH_4 = 6.02×10^{23} molecules CH_4

7. (a) 6.02×10^{23} atoms S = 1 mol S
   (b) 6.02×10^{23} molecules SO_2 = 1 mol SO_2

9. (a) $$0.100\ \cancel{\text{mol Mg}} \times \frac{6.02 \times 10^{23}\text{ atoms Mg}}{1\ \cancel{\text{mol Mg}}} = 6.02 \times 10^{22}\text{ atoms Mg}$$

   (b) $$0.200\ \cancel{\text{mol Cl}_2} \times \frac{6.02 \times 10^{23}\text{ molecules Cl}_2}{1\ \cancel{\text{mol Cl}_2}} = 1.20 \times 10^{23}\text{ molecules Cl}_2$$

   (c) $$0.300\ \cancel{\text{mol MgCl}_2} \times \frac{6.02 \times 10^{23}\text{ formula units MgCl}_2}{1\ \cancel{\text{mol MgCl}_2}}$$

   $$= 1.81 \times 10^{23}\text{ formula units MgCl}_2$$

11. (a) $2.50 \times 10^{22}\ \cancel{\text{atoms Fe}} \times \dfrac{1\text{ mol Fe}}{6.02 \times 10^{23}\ \cancel{\text{atoms Fe}}} = 0.0415\text{ mol Fe}$

(b) $5.00 \times 10^{23}\ \cancel{\text{molecules SO}_3} \times \dfrac{1\text{ mol SO}_3}{6.02 \times 10^{23}\ \cancel{\text{molecules SO}_3}} = 0.831\text{ mol SO}_3$

(c) $7.50 \times 10^{24}\ \cancel{\text{formula units FeSO}_4} \times \dfrac{1\text{ mol FeSO}_4}{6.02 \times 10^{23}\ \cancel{\text{formula units FeSO}_4}}$

$= 12.5\text{ mol FeSO}_4$

Section 8.3 *Molar Mass*

13.

	Element	Molar Mass
(a)	Al	26.98 g/mol
(b)	Si	28.09 g/mol
(c)	As	74.92 g/mol
(d)	S	32.07 g/mol

15.

	Compound	Molar Mass
(a)	CaS	40.08 g Ca + 32.07 g S = 72.15 g/mol
(b)	$CaSO_4$	40.08 g Ca + 32.07 g S + 4(16.00 g O) = 136.15 g/mol
(c)	Cl_2O_5	2(35.45 g) + 5(16.00 g) = 150.90 g/mol
(d)	$C_3H_5(OH)_3$	3(12.01 g) + 5(1.01 g) + 3(16.00 g + 1.01 g) = 92.11 g/mol

Section 8.4 *Mole Calculations II*

17. (a) MM of Hg = 200.59 g/mol

$2.95 \times 10^{23}\ \cancel{\text{atoms Hg}} \times \dfrac{1\ \cancel{\text{mol Hg}}}{6.02 \times 10^{23}\ \cancel{\text{atoms Hg}}} \times \dfrac{200.59\text{ g Hg}}{1\ \cancel{\text{mol Hg}}} = 98.3\text{ g Hg}$

(b) MM of N_2 = 2(14.01 g N) = 28.02 g/mol

$1.16 \times 10^{22}\ \cancel{\text{molecules N}_2} \times \dfrac{1\ \cancel{\text{mol N}_2}}{6.02 \times 10^{23}\ \cancel{\text{molecules N}_2}} \times \dfrac{28.02\text{ g N}_2}{1\ \cancel{\text{mol N}_2}}$

$= 0.540\text{ g N}_2$

(c) MM of $BaCl_2$ = 137.33 g Ba + 2(35.45 g Cl) = 208.23 g/mol

$$5.05 \times 10^{21} \text{ formula units } BaCl_2 \times \frac{1 \text{ mol } BaCl_2}{6.02 \times 10^{23} \text{ formula units } BaCl_2}$$

$$\times \frac{208.23 \text{ g } BaCl_2}{1 \text{ mol } BaCl_2} = 1.75 \text{ g } BaCl_2$$

19. (a) MM of K = 39.10 g/mol

$$1.50 \text{ g K} \times \frac{1 \text{ mol K}}{39.10 \text{ g K}} \times \frac{6.02 \times 10^{23} \text{ atoms K}}{1 \text{ mol K}} = 2.31 \times 10^{22} \text{ atoms K}$$

(b) MM of O_2 = 2(16.00 g O) = 32.00 g/mol

$$0.470 \text{ g } O_2 \times \frac{1 \text{ mol } O_2}{32.00 \text{ g } O_2} \times \frac{6.02 \times 10^{23} \text{ molecules } O_2}{1 \text{ mol } O_2}$$

$$= 8.84 \times 10^{21} \text{ molecules } O_2$$

(c) MM of $AgClO_3$ = 107.87 g Ag + 35.45 g Cl + 3(16.00 g O) = 191.32 g/mol

$$0.555 \text{ g } AgClO_3 \times \frac{1 \text{ mol } AgClO_3}{191.32 \text{ g } AgClO_3} \times \frac{6.02 \times 10^{23} \text{ formula units } AgClO_3}{1 \text{ mol } AgClO_3}$$

$$= 1.75 \times 10^{21} \text{ formula units } AgClO_3$$

21. (a) $\frac{6.94 \text{ g}}{1 \text{ mol Li}} \times \frac{1 \text{ mol Li}}{6.02 \times 10^{23} \text{ atoms}} = 1.15 \times 10^{-23} \text{ g/atom}$

(b) $\frac{28.06 \text{ g}}{1 \text{ mol Si}} \times \frac{1 \text{ mol Si}}{6.02 \times 10^{23} \text{ atoms}} = 4.66 \times 10^{-23} \text{ g/atom}$

(c) MM of SO_2 = 32.07 g S + 2(16.00 g O) = 64.07 g/mol

$$\frac{64.07 \text{ g}}{1 \text{ mol } SO_2} \times \frac{1 \text{ mol } SO_2}{6.02 \times 10^{23} \text{ molecules}} = 1.06 \times 10^{-22} \text{ g/molecule}$$

(d) MM of N_2O = 2(14.01 g N) + 16.00 g O = 44.02 g/mol

$$\frac{44.02 \text{ g}}{1 \text{ mol } N_2O} \times \frac{1 \text{ mol } N_2O}{6.02 \times 10^{23} \text{ molecules}} = 7.31 \times 10^{-23} \text{ g/molecule}$$

Section 8.5 *Molar Volume*

23. Standard conditions: 0 °C and 1 atmosphere (1 atm)

25. (a) MM of Ar = 39.95 g/mol

Density of Ar (at STP): $\frac{39.95\text{ g}}{1\text{ }\cancel{\text{mol}}} \times \frac{1\text{ }\cancel{\text{mol}}}{22.4\text{ L}} = 1.78\text{ g/L}$

(b) MM of F_2 = 2(19.00 g F) = 38.00 g/mol

Density of F_2 (at STP): $\frac{38.00\text{ g}}{1\text{ }\cancel{\text{mol}}} \times \frac{1\text{ }\cancel{\text{mol}}}{22.4\text{ L}} = 1.70\text{ g/L}$

(c) MM of CH_4 = 12.01 g C + 4(1.01 g H) = 16.05 g/mol

Density of CH_4 (at STP): $\frac{16.05\text{ g}}{1\text{ }\cancel{\text{mol}}} \times \frac{1\text{ }\cancel{\text{mol}}}{22.4\text{ L}} = 0.717\text{ g/L}$

(d) MM of C_3H_8 = 3(12.01 g C) + 8(1.01 g H) = 44.11 g/mol

Density of C_3H_8 (at STP): $\frac{44.11\text{ g}}{1\text{ }\cancel{\text{mol}}} \times \frac{1\text{ }\cancel{\text{mol}}}{22.4\text{ L}} = 1.97\text{ g/L}$

27. (a) MM of oxygen: $\frac{1.43\text{ g}}{1\text{ }\cancel{\text{L}}} \times \frac{22.4\text{ }\cancel{\text{L}}}{\text{mol}} = 32.0\text{ g/mol}$

(b) MM of phosphine: $\frac{1.52\text{ g}}{1\text{ }\cancel{\text{L}}} \times \frac{22.4\text{ }\cancel{\text{L}}}{\text{mol}} = 34.0\text{ g/mol}$

(c) MM of nitrous oxide: $\frac{1.97\text{ g}}{1\text{ }\cancel{\text{L}}} \times \frac{22.4\text{ }\cancel{\text{L}}}{\text{mol}} = 44.1\text{ g/mol}$

(d) MM of Freon-12: $\frac{5.40\text{ g}}{1\text{ }\cancel{\text{L}}} \times \frac{22.4\text{ }\cancel{\text{L}}}{\text{mol}} = 121\text{ g/mol}$

29.

Gas	Molecules	Mass	Volume at STP
nitrogen, N_2	6.02×10^{23}	28.02 g	22.4 L
oxygen, O_2	6.02×10^{23}	32.00 g	22.4 L
nitrogen dioxide, NO_2	6.02×10^{23}	46.01 g	22.4 L

Section 8.6 *Mole Calculations III*

31. (a) MM of CO = 12.01 g C + 16.00 g O = 28.01 g/mol

$$0.150 \text{ } \cancel{\text{g CO}} \times \frac{1 \text{ } \cancel{\text{mol CO}}}{28.01 \text{ } \cancel{\text{g CO}}} \times \frac{6.02 \times 10^{23} \text{ molecules } H_2}{1 \text{ } \cancel{\text{mol CO}}}$$

$$= 3.22 \times 10^{21} \text{ molecules CO}$$

(b) $$100.0 \text{ } \cancel{\text{mL } H_2} \times \frac{1 \text{ } \cancel{\text{L } H_2}}{1000 \text{ } \cancel{\text{mL } H_2}} \times \frac{1 \text{ } \cancel{\text{mol } H_2}}{22.4 \text{ } \cancel{\text{L } H_2}} \times \frac{6.02 \times 10^{23} \text{ molecules } H_2}{1 \text{ } \cancel{\text{mol } H_2}}$$

$$= 2.69 \times 10^{21} \text{ molecules } H_2$$

33. (a) MM of H_2S = 2(1.01 g H) + 32.07 g S = 34.09 g/mol

$$1.05 \text{ } \cancel{\text{L } H_2S} \times \frac{1 \text{ } \cancel{\text{mol } H_2S}}{22.4 \text{ } \cancel{\text{L } H_2S}} \times \frac{34.09 \text{ g } H_2S}{1 \text{ } \cancel{\text{mol } H_2S}} = 1.60 \text{ g } H_2S$$

(b) MM of C_3H_8 = 3(12.01 g C) + 8(1.01 g H) = 44.11 g/mol

$$5.42 \times 10^{22} \text{ } \cancel{\text{molecules } C_3H_8} \times \frac{1 \text{ } \cancel{\text{mol } C_3H_8}}{6.02 \times 10^{23} \text{ } \cancel{\text{molecules } C_3H_8}} \times \frac{44.11 \text{ g } C_3H_8}{1 \text{ } \cancel{\text{mol } C_3H_8}}$$

$$= 3.97 \text{ g } C_3H_8$$

35. (a) MM of He = 4.00 g/mol

$$0.250 \text{ } \cancel{\text{g He}} \times \frac{1 \text{ } \cancel{\text{mol He}}}{4.00 \text{ } \cancel{\text{g He}}} \times \frac{22.4 \text{ L He}}{1 \text{ } \cancel{\text{mol He}}} = 1.40 \text{ L He}$$

(b) $$2.22 \times 10^{22} \text{ } \cancel{\text{molecules } CH_4} \times \frac{1 \text{ } \cancel{\text{mol } CH_4}}{6.02 \times 10^{23} \text{ } \cancel{\text{molecules } CH_4}} \times \frac{22.4 \text{ L } CH_4}{1 \text{ } \cancel{\text{mol } CH_4}}$$

$$= 0.826 \text{ L } CH_4$$

37.

Gas	Molecules	Mass	Volume at STP
methane, CH_4	1.50×10^{23}	4.00 g	5.58 L
ethane, C_2H_6	1.50×10^{23}	7.50 g	5.58 L
propane, C_3H_8	1.50×10^{23}	11.0 g	5.58 L

Section 8.7 *Percent Composition*

39. The law of definite composition states that a compound always contains the same elements in the same proportion by mass. Thus, the percent sodium in a kilogram of salt 39.34%.

41. MM of $C_{12}H_{22}O_{11}$ $= 12(12.01\text{ g C}) + 22(1.01\text{ g H}) + 11(16.00\text{ g O})$

$$= 144.1\text{ g C} + 22.2\text{ g H} + 176.0\text{ g O}$$

$$= 342.3\text{ g/mol}$$

$$\frac{144.1\text{ g C}}{342.3\text{ g }C_{12}H_{22}O_{11}} \times 100\% = 42.10\%\text{ C}$$

$$\frac{22.2\text{ g H}}{342.3\text{ g }C_{12}H_{22}O_{11}} \times 100\% = 6.49\%\text{ H}$$

$$\frac{176.0\text{ g O}}{342.3\text{ g }C_{12}H_{22}O_{11}} \times 100\% = 51.42\%\text{ O}$$

43. MM of $C_5H_{11}NSO_2$

$$= 5(12.01\text{ g C}) + 11(1.01\text{ g H}) + 14.01\text{ g N} + 32.07\text{ g S} + 2(16.00\text{ g O})$$

$$= 60.05\text{ g C} + 11.11\text{ g H} + 14.01\text{ g N} + 32.07\text{ g S} + 32.00\text{ g O}$$

$$= 149.24\text{ g/mol}$$

$$\frac{60.05\text{ g C}}{149.24\text{ g }C_5H_{11}NSO_2} \times 100\% = 40.24\%\text{ C}$$

$$\frac{11.11\text{ g H}}{149.24\text{ g }C_5H_{11}NSO_2} \times 100\% = 7.444\%\text{ H}$$

$$\frac{14.01\text{ g N}}{149.24\text{ g }C_5H_{11}NSO_2} \times 100\% = 9.388\%\text{ N}$$

$$\frac{32.07\text{ g S}}{149.24\text{ g }C_5H_{11}NSO_2} \times 100\% = 21.49\%\text{ S}$$

$$\frac{32.00\text{ g O}}{149.24\text{ g }C_5H_{11}NSO_2} \times 100\% = 21.44\%\text{ O}$$

45. MM of $HgNa_2C_{20}H_8Br_2O_6$

$$= 200.59 \text{ g Hg} + 2(22.99 \text{ g Na}) + 20(12.01 \text{ g C}) + 8(1.01 \text{ g H}) + 2(79.90 \text{ g Br}) + 6(16.00 \text{ g O})$$

$$= 200.59 \text{ g Hg} + 45.98 \text{ g Na} + 240.20 \text{ g C} + 8.08 \text{ g H} + 159.80 \text{ g Br} + 96.00 \text{ g O}$$

$$= 750.65 \text{ g/mol}$$

$$\frac{200.59 \text{ g Hg}}{750.65 \text{ g } HgNa_2C_{20}H_8Br_2O_6} \times 100\% = 26.722\% \text{ Hg}$$

$$\frac{45.98 \text{ g Na}}{750.65 \text{ g } HgNa_2C_{20}H_8Br_2O_6} \times 100\% = 6.125\% \text{ Na}$$

$$\frac{240.20 \text{ g C}}{750.65 \text{ g } HgNa_2C_{20}H_8Br_2O_6} \times 100\% = 31.999\% \text{ C}$$

$$\frac{8.08 \text{ g H}}{750.65 \text{ g } HgNa_2C_{20}H_8Br_2O_6} \times 100\% = 1.08\% \text{ H}$$

$$\frac{159.80 \text{ g Br}}{750.65 \text{ g } HgNa_2C_{20}H_8Br_2O_6} \times 100\% = 21.288\% \text{ Br}$$

$$\frac{96.00 \text{ g O}}{750.65 \text{ g } HgNa_2C_{20}H_8Br_2O_6} \times 100\% = 12.79\% \text{ O}$$

Section 8.8 *Empirical Formula*

47. $$1.115 \text{ } \cancel{\text{g Co}} \times \frac{1 \text{ mol Co}}{58.93 \text{ } \cancel{\text{g Co}}} = 0.0189 \text{ mol Co}$$

$$1.722 \text{ g } Co_xS_y - 1.115 \text{ g Co} = 0.607 \text{ g S}$$

$$0.607 \text{ } \cancel{\text{g S}} \times \frac{1 \text{ mol S}}{32.07 \text{ } \cancel{\text{g S}}} = 0.0189 \text{ mol S}$$

$$\frac{Co_{0.0189}}{0.0189} \frac{S_{0.0189}}{0.0189} = Co_{1.00}S_{1.00}$$

The empirical formula is CoS.

49. $1.925\ \cancel{g\ Cu} \times \frac{1\ mol\ Cu}{63.55\ \cancel{g\ Cu}} = 0.03029\ mol\ Cu$

$2.410\ g\ Cu_xO_y - 1.925\ g\ Cu = 0.485\ g\ O$

$0.485\ \cancel{g\ O} \times \frac{1\ mol\ O}{16.00\ \cancel{g\ O}} = 0.0303\ mol\ O$

$\frac{Cu_{0.03029}}{0.0303}\frac{O_{0.0303}}{0.0303} = Cu_{1.00}O_{1.00}$ The empirical formula is CuO.

51. $0.500\ \cancel{g\ Sn} \times \frac{1\ mol\ Sn}{118.71\ \cancel{g\ Sn}} = 0.00421\ mol\ Sn$

$0.635\ g\ Sn_xO_y - 0.500\ g\ Sn = 0.135\ g\ O$

$0.135\ \cancel{g\ O} \times \frac{1\ mol\ O}{16.00\ \cancel{g\ O}} = 0.00844\ mol\ O$

$\frac{Sn_{0.00421}}{0.00421}\frac{O_{0.00844}}{0.00421} = Sn_{1.00}O_{2.00}$ The empirical formula is SnO_2.

53. $64.1\ \cancel{g\ Cu} \times \frac{1\ mol\ Cu}{63.55\ \cancel{g\ Cu}} = 1.01\ mol\ Cu$

$35.9\ \cancel{g\ Cl} \times \frac{1\ mol\ Cl}{35.45\ \cancel{g\ Cl}} = 1.01\ mol\ Cl$

$\frac{Cu_{1.01}}{1.01}\frac{Cl_{1.01}}{1.01} = Cu_{1.00}Cl_{1.00}$ The empirical formula is CuCl.

55. $42.6\ \cancel{g\ Sn} \times \frac{1\ mol\ Sn}{118.71\ \cancel{g\ Sn}} = 0.359\ mol\ Sn$

$57.4\ \cancel{g\ Br} \times \frac{1\ mol\ Br}{79.90\ \cancel{g\ Br}} = 0.718\ mol\ Br$

$\frac{Sn_{0.359}}{0.359}\frac{Br_{0.718}}{0.359} = Sn_{1.00}Br_{2.00}$ The empirical formula is $SnBr_2$.

57. $18.25\ \text{g C} \times \dfrac{1\ \text{mol C}}{12.01\ \text{g C}} = 1.52\ \text{mol C}$

$0.77\ \text{g H} \times \dfrac{1\ \text{mol H}}{1.01\ \text{g H}} = 0.76\ \text{mol H}$

$80.99\ \text{g Cl} \times \dfrac{1\ \text{mol Cl}}{35.45\ \text{g Cl}} = 2.28\ \text{mol Cl}$

$\dfrac{\text{C}\,1.52}{0.76}\ \dfrac{\text{H}\,0.76}{0.76}\ \dfrac{\text{Cl}\,2.28}{0.76} = C_{2.0}H_{1.0}Cl_{3.0}$ The empirical formula is C_2HCl_3.

Section 8.9 *Molecular Formula*

59. MM of $C_9H_8O_4$ $= 9(12.01\ \text{g C}) + 8(1.01\ \text{g H}) + 4(16.00\ \text{g O})$
$= 108.19\ \text{g C} + 8.08\ \text{g H} + 64.00\ \text{g O}$
$= 180.17\ \text{g/mol}$

Aspirin: $\dfrac{(C_9H_8O_4)_n}{C_9H_8O_4} = \dfrac{180\ \text{g/mol}}{180.17\ \text{g/mol}}$ $n \approx 1$

The molecular formula of aspirin is $(C_9H_8O_4)_1$ or $C_9H_8O_4$.

61. MM of $C_3H_5O_2$ $= 3(12.01\ \text{g C}) + 5(1.01\ \text{g H}) + 2(16.00\ \text{g O})$
$= 36.03\ \text{g C} + 5.05\ \text{g H} + 32.00\ \text{g O}$
$= 73.08\ \text{g/mol}$

Adipic acid: $\dfrac{(C_3H_5O_2)_n}{C_3H_5O_2} = \dfrac{147\ \text{g/mol}}{73.08\ \text{g/mol}}$ $n \approx 2$

The molecular formula of adipic acid is $(C_3H_5O_2)_2$ or $C_6H_{10}O_4$.

63. Empirical Formula

$38.7\ \text{g C} \times \dfrac{1\ \text{mol C}}{12.01\ \text{g C}} = 3.22\ \text{mol C}$

$9.74\ \text{g H} \times \dfrac{1\ \text{mol H}}{1.01\ \text{g H}} = 9.64\ \text{mol H}$

$51.6\ \text{g O} \times \dfrac{1\ \text{mol O}}{16.00\ \text{g O}} = 3.23\ \text{mol O}$

$\dfrac{\text{C}\,3.22}{3.22}\ \dfrac{\text{H}\,9.64}{3.22}\ \dfrac{\text{O}\,3.23}{3.22} = C_{1.00}H_{2.99}O_{1.00}$ The empirical formula is CH_3O.

Molecular Formula

MM of CH_3O = 12.01 g C + 3(1.01 g H) + 16.00 g O
= 12.01 g C + 3.03 g H + 16.00 g O
= 31.04 g/mol

Ethylene glycol: $\frac{(CH_3O)_n}{CH_3O} = \frac{62 \text{ g/mol}}{31.04 \text{ g/mol}}$ $\quad n \approx 2$

The molecular formula of ethylene glycol is $(CH_3O)_2$ or $C_2H_6O_2$.

65. Empirical Formula

$$24.8 \text{ g C} \times \frac{1 \text{ mol C}}{12.01 \text{ g C}} = 2.06 \text{ mol C}$$

$$2.08 \text{ g H} \times \frac{1 \text{ mol H}}{1.01 \text{ g H}} = 2.06 \text{ mol H}$$

$$73.1 \text{ g Cl} \times \frac{1 \text{ mol Cl}}{35.45 \text{ g Cl}} = 2.06 \text{ mol Cl}$$

$$\frac{C\,2.06}{2.06}\,\frac{H\,2.06}{2.06}\,\frac{Cl\,2.06}{2.06} = C_{1.00}H_{1.00}Cl_{1.00}$$

The empirical formula is CHCl.

Molecular Formula

MM of CHCl = 12.01 g C + 1.01 g H + 35.45 g Cl
= 48.47 g/mol

Lindane: $\frac{(CHCl)_n}{CHCl} = \frac{290 \text{ g/mol}}{48.47 \text{ g/mol}}$ $\quad n \approx 6$

The molecular formula of lindane is $(CHCl)_6$ or $C_6H_6Cl_6$.

67. Empirical Formula

$$74.0 \text{ g C} \times \frac{1 \text{ mol C}}{12.01 \text{ g C}} = 6.16 \text{ mol C}$$

$$8.70 \text{ g H} \times \frac{1 \text{ mol H}}{1.01 \text{ g H}} = 8.61 \text{ mol H}$$

$$17.3 \text{ g N} \times \frac{1 \text{ mol N}}{14.01 \text{ g N}} = 1.23 \text{ mol N}$$

$$\frac{C\,6.16}{1.23}\,\frac{H\,8.61}{1.23}\,\frac{N\,1.23}{1.23} = C_{5.01}H_{7.00}N_{1.00}$$

The empirical formula is C_5H_7N.

Molecular Formula

$$\text{MM of } C_5H_7N = 5(12.01 \text{ g C}) + 7(1.01 \text{ g H}) + 14.01 \text{ g N}$$
$$= 60.05 \text{ g C} + 7.07 \text{ g H} + 14.01 \text{ g N}$$
$$= 81.13 \text{ g/mol}$$

Nicotine: $\dfrac{(C_5H_7N)_n}{C_5H_7N} = \dfrac{160 \text{ g/mol}}{81.13 \text{ g/mol}} \qquad n \approx 2$

The molecular formula of nicotine is $(C_5H_7N)_2$ or $C_{10}H_{14}N_2$.

General Exercises

69. $0.500 \text{ }\cancel{\text{g Ge}} \times \dfrac{1 \text{ mol Ge}}{72.61 \text{ }\cancel{\text{g Ge}}} = 0.00689 \text{ mol Ge}$

$1.456 \text{ g } Ge_xCl_y - 0.500 \text{ g Ge} = 0.956 \text{ g Cl}$

$0.956 \text{ }\cancel{\text{g Cl}} \times \dfrac{1 \text{ mol Cl}}{35.45 \text{ }\cancel{\text{g Cl}}} = 0.0270 \text{ mol Cl}$

$\text{Ge}\,\dfrac{0.00689}{0.00689}\ \ \text{Cl}\,\dfrac{0.0270}{0.00689} = Ge_{1.00}Cl_{3.92}$ The empirical formula is $GeCl_4$.

71. $5 \text{ }\cancel{\text{g Ni}} \times \dfrac{1 \text{ }\cancel{\text{mol Ni}}}{58.69 \text{ }\cancel{\text{g Ni}}} \times \dfrac{6.02 \times 10^{23} \text{ atoms Ni}}{1 \text{ }\cancel{\text{mol Ni}}} = 5 \times 10^{22} \text{ atoms Ni}$

5×10^{22} atoms Ni $>>>$ 1×10^{15} red blood cells

Thus, the number of Ni atoms in a 5-g nickel coin is many times greater than the number of red blood cells in 50,000 people!

73. $1 \text{ }\cancel{\text{mol furry moles}} \times \dfrac{6.02 \times 10^{23} \text{ }\cancel{\text{furry moles}}}{1 \text{ }\cancel{\text{mol furry moles}}} \times \dfrac{100 \text{ }\cancel{\text{g}}}{1 \text{ }\cancel{\text{furry mole}}} \times \dfrac{1 \text{ kg}}{1000 \text{ }\cancel{\text{g}}}$

$$= 6 \times 10^{22} \text{ kg}$$

6×10^{24} kg (Earth) $>>$ 6×10^{22} kg (1 mol furry moles)
Thus, the Earth weighs about 100 times more than a mole of moles!

75. MM of $C_{12}H_{22}O_{11}$ $= 12(12.01\text{ g C}) + 22(1.01\text{ g H}) + 11(16.00\text{ g O})$

$= 144.12\text{ g C} + 22.22\text{ g H} + 176.00\text{ g O}$

$= 342.34\text{ g/mol}$

$$1.00\ \cancel{\text{g } C_{12}H_{22}O_{11}} \times \frac{1\ \cancel{\text{mol}}}{342.34\ \cancel{\text{g}}} \times \frac{12\ \cancel{\text{mol C}}}{1\ \cancel{\text{mol } C_{12}H_{22}O_{11}}} \times \frac{6.02 \times 10^{23}\text{ atoms C}}{1\ \cancel{\text{mol C}}}$$

$$= 2.11 \times 10^{22}\text{ atoms C}$$

77. $$10.0\ \cancel{\text{g Cu}} \times \frac{1\ \cancel{\text{mol Cu}}}{63.55\ \cancel{\text{g Cu}}} \times \frac{1\ \cancel{\text{mol } Cu_2O}}{2\ \cancel{\text{mol Cu}}} \times \frac{143.10\text{ g } Cu_2O}{1\ \cancel{\text{mol } Cu_2O}} = 11.3\text{ g } Cu_2O$$

Challenge Exercises

79. $$1\ \cancel{\text{molecule } H_2O} \times \frac{1\ \cancel{\text{mol } H_2O}}{6.02 \times 10^{23}\ \cancel{\text{molecules } H_2O}} \times \frac{18.02\ \cancel{\text{g } H_2O}}{1\ \cancel{\text{mol } H_2O}} \times \frac{1\text{ cm}^3\ H_2O}{1.00\ \cancel{\text{g } H_2O}}$$

$$= 2.99 \times 10^{-23}\text{ cm}^3\ H_2O$$

81. $$\frac{1\text{ atom Cu}}{0.0118\ \cancel{\text{nm}^3}} \times \left(\frac{1 \times 10^9\ \cancel{\text{nm}}}{1\ \cancel{\text{m}}}\right)^3 \times \left(\frac{1\ \cancel{\text{m}}}{100\ \cancel{\text{cm}}}\right)^3 \times \frac{1\ \cancel{\text{cm}^3}}{8.92\ \cancel{\text{g Cu}}} \times \frac{63.55\ \cancel{\text{g Cu}}}{1\text{ mol Cu}}$$

$$= \frac{6.04 \times 10^{23}\text{ atoms Cu}}{1\text{ mol Cu}}$$

CHAPTER

9

Chemical Equation Calculations

Section 9.1 *Interpreting a Chemical Equation*

1. General Equation: A + 2 B → C + 2 D

(a) $1\ \cancel{\text{mol A}} \times \dfrac{1\text{ mol C}}{1\ \cancel{\text{mol A}}} = 1\text{ mol C}$

(b) $2\ \cancel{\text{L D}} \times \dfrac{2\text{ L B}}{2\ \cancel{\text{L D}}} = 2\text{ L B}$

3. General Equation: 2 A + 3 B → C
(a) conservation of mass law: 1.00 g A + 1.50 g B = 2.50 g C
(b) conservation of mass law: 5.00 g C - 2.00 g A = 3.00 g B

5. (a) $2\ N_2(g) + 3\ O_2(g) \rightarrow 2\ N_2O_3(g)$

MM of N_2 = 2(14.01 g N) = 28.02 g/mol
MM of O_2 = 2(16.00 g O) = 32.00 g/mol
MM of N_2O_3 = 2(14.01 g N) + 3(16.00 g O) = 76.02 g/mol

2(28.02 g N_2) + 3(32.00 g O_2) → 2(76.02 g N_2O_3)
56.04 g N_2 + 96.00 g O_2 → 152.04 g N_2O_3
152.04 g → 152.04 g

(b) $2\ N_2(s) + 5\ O_2(g) \rightarrow 2\ N_2O_5(s)$

MM of N_2 = 2(14.01 g N) = 28.02 g/mol
MM of O_2 = 2(16.00 g O) = 32.00 g/mol
MM of N_2O_5 = 2(14.01 g N) + 5(16.00 g O) = 108.02 g/mol

2(28.02 g N2) + 5(32.00 g O_2) → 2(108.02 g N_2O_5)
56.04 g N_2 + 160.00 g O_2 → 216.04 g N_2O_5
216.04 g → 216.04 g

Section 9.2 *Mole–Mole Relationships*

7. $2\ H_2O_2(l) \rightarrow 2\ H_2O(l) + O_2(g)$

Moles of O_2 produced:

$$5.00\ \cancel{\text{mol } H_2O_2} \times \frac{1\ \text{mol } O_2}{2\ \cancel{\text{mol } H_2O_2}} = 2.50\ \text{mol } O_2$$

9. $2\ LiClO_3(s) \rightarrow 2\ LiCl(s) + O_2(g)$

Moles of O_2 produced:

$$5.00\ \cancel{\text{mol } LiClO_3} \times \frac{1\ \text{mol } O_2}{2\ \cancel{\text{mol } LiClO_3}} = 2.50\ \text{mol } O_2$$

11. $2\ Fe(s) + 3\ Cl_2(g) \rightarrow 2\ FeCl_3(s)$

Moles of Cl_2 that react:

$$0.100\ \cancel{\text{mol Fe}} \times \frac{3\ \text{mol } Cl_2}{2\ \cancel{\text{mol Fe}}} = 0.150\ \text{mol } Cl_2$$

Moles of $FeCl_3$ produced:

$$0.100\ \cancel{\text{mol Fe}} \times \frac{2\ \text{mol } FeCl_3}{2\ \cancel{\text{mol Fe}}} = 0.100\ \text{mol } FeCl_3$$

13. $C_3H_8(g) + 5\ O_2(g) \rightarrow 3\ CO_2(g) + 4\ H_2O(g)$

Moles of $C_3H_8(g)$ that react:

$$2.50\ \cancel{\text{mol } O_2} \times \frac{1\ \text{mol } C_3H_8}{5\ \cancel{\text{mol } O_2}} = 0.500\ \text{mol } C_3H_8$$

Moles of $CO_2(g)$ produced:

$$2.50\ \cancel{\text{mol } O_2} \times \frac{3\ \text{mol } CO_2}{5\ \cancel{\text{mol } O_2}} = 1.50\ \text{mol } CO_2$$

Section 9.3 *Types of Stoichiometry Problems*

15.

Given	Unknown	Type of Stoichiometry
mass	mass	mass–mass problem

17.

Given	Unknown	Type of Stoichiometry
mass	volume	mass–volume problem

19.

Given	Unknown	Type of Stoichiometry
volume	volume	volume–volume problem

Section 9.4 *Mass–Mass Problems*

21. $2\,Zn(s) + O_2(g) \rightarrow 2\,ZnO(s)$

MM of Zn = 65.39 g/mol
MM of ZnO = 81.39 g/mol

$$1.00\ \cancel{g\ Zn} \times \frac{1\ \cancel{mol\ Zn}}{65.39\ \cancel{g\ Zn}} \times \frac{2\ \cancel{mol\ ZnO}}{2\ \cancel{mol\ Zn}} \times \frac{81.39\ g\ ZnO}{1\ \cancel{mol\ ZnO}} = 1.24\ g\ ZnO$$

23. $2\,Bi(s) + 3\,Cl_2(g) \rightarrow 2\,BiCl_3(s)$

MM of Bi = 208.98 g/mol
MM of $BiCl_3$ = 315.33 g/mol

$$3.45\ \cancel{g\ Bi} \times \frac{1\ \cancel{mol\ Bi}}{208.98\ \cancel{g\ Bi}} \times \frac{2\ \cancel{mol\ BiCl_3}}{2\ \cancel{mol\ Bi}} \times \frac{315.33\ g\ BiCl_3}{1\ \cancel{mol\ BiCl_3}} = 5.21\ g\ BiCl_3$$

25. $Cu(s) + 2\,AgNO_3(aq) \rightarrow Cu(NO_3)_2(aq) + 2\,Ag(s)$

MM of Cu = 63.55 g/mol
MM of Ag = 107.87 g/mol

$$1.00\ \cancel{g\ Cu} \times \frac{1\ \cancel{mol\ Cu}}{63.55\ \cancel{g\ Cu}} \times \frac{2\ \cancel{mol\ Ag}}{1\ \cancel{mol\ Cu}} \times \frac{107.87\ g\ Ag}{1\ \cancel{mol\ Ag}} = 3.39\ g\ Ag$$

27. $2\,Co(s) + 3\,HgCl_2(aq) \rightarrow 2\,CoCl_3(aq) + 3\,Hg(l)$

MM of Co = 58.93 g/mol
MM of Hg = 200.59 g/mol

$$1.25\ \cancel{g\ Co} \times \frac{1\ \cancel{mol\ Co}}{58.93\ \cancel{g\ Co}} \times \frac{3\ \cancel{mol\ Hg}}{2\ \cancel{mol\ Co}} \times \frac{200.59\ g\ Hg}{1\ \cancel{mol\ Hg}} = 6.38\ g\ Hg$$

29. $2\,Na_3PO_4(aq) + 3\,Ca(OH)_2(aq) \rightarrow Ca_3(PO_4)_2(s) + 6\,NaOH(aq)$

MM of Na_3PO_4 = 163.94 g/mol
MM of $Ca_3(PO_4)_2$ = 310.18 g/mol

$$1.78\ \cancel{g\ Na_3PO_4} \times \frac{1\ \cancel{mol\ Na_3PO_4}}{163.94\ \cancel{g\ Na_3PO_4}} \times \frac{1\ \cancel{mol\ Ca_3(PO_4)_2}}{2\ \cancel{mol\ Na_3PO_4}} \times \frac{310.18\ g\ Ca_3(PO_4)_2}{1\ \cancel{mol\ Ca_3(PO_4)_2}}$$

$$= 1.68\ g\ Ca_3(PO_4)_2$$

Section 9.5 *Mass–Volume Problems*

31. $2\,NaHCO_3(s) \rightarrow Na_2CO_3(s) + H_2O(g) + CO_2(g)$

MM of $NaHCO_3$ = 84.01 g/mol

$$1.75\ \text{g NaHCO}_3 \times \frac{1\ \text{mol NaHCO}_3}{84.01\ \text{g NaHCO}_3} \times \frac{1\ \text{mol CO}_2}{2\ \text{mol NaHCO}_3} \times \frac{22.4\ \text{L CO}_2}{1\ \text{mol CO}_2} \times \frac{1000\ \text{mL CO}_2}{1\ \text{L CO}_2} = 233\ \text{mL CO}_2$$

33. $Fe_2(CO_3)_3(s) \xrightarrow{\Delta} Fe_2O_3(s) + 3\,CO_2(g)$

MM of $Fe_2(CO_3)_3$ = 291.73 g/mol

$$5.00\ \text{g Fe}_2(\text{CO}_3)_3 \times \frac{1\ \text{mol Fe}_2(\text{CO}_3)_3}{291.73\ \text{g Fe}_2(\text{CO}_3)_3} \times \frac{3\ \text{mol CO}_2}{1\ \text{mol Fe}_2(\text{CO}_3)_3} \times \frac{22.4\ \text{L CO}_2}{1\ \text{mol CO}_2} \times \frac{1000\ \text{mL CO}_2}{1\ \text{L CO}_2} = 1150\ \text{mL CO}_2$$

35. $2\,HgO(s) \rightarrow 2\,Hg(l) + O_2(g)$

MM of HgO = 216.59 g/mol

$$2.50\ \text{g HgO} \times \frac{1\ \text{mol HgO}}{216.59\ \text{g HgO}} \times \frac{1\ \text{mol O}_2}{2\ \text{mol HgO}} \times \frac{22.4\ \text{L O}_2}{1\ \text{mol O}_2} \times \frac{1000\ \text{mL O}_2}{1\ \text{L O}_2} = 129\ \text{mL O}_2$$

37. $2\,H_2O_2(l) \rightarrow 2\,H_2O(l) + O_2(g)$

MM of H_2O_2 = 34.02 g/mol

$$1.55\ \text{g H}_2\text{O}_2 \times \frac{1\ \text{mol H}_2\text{O}_2}{34.02\ \text{g H}_2\text{O}_2} \times \frac{1\ \text{mol O}_2}{2\ \text{mol H}_2\text{O}_2} \times \frac{22.4\ \text{L O}_2}{1\ \text{mol O}_2} \times \frac{1000\ \text{mL O}_2}{1\ \text{L O}_2} = 510\ \text{mL O}_2$$

Section 9.6 *Volume–Volume Problems*

39. $3\,H_2(g) + N_2(g) \rightarrow 2\,NH_3(g)$

$$50.0\ \text{L H}_2 \times \frac{1\ \text{L N}_2}{3\ \text{L H}_2} = 16.7\ \text{L N}_2$$

41. $2\,CO(g) + O_2(g) \rightarrow 2\,CO_2(g)$

$$10.0\ \text{mL CO} \times \frac{1\ \text{mL O}_2}{2\ \text{mL CO}} = 5.00\ \text{mL O}_2$$

43. $2\,Cl_2(g) + 3\,O_2(g) \rightarrow 2\,Cl_2O_3(g)$

$$1.75\ \cancel{L\,Cl_2O_3} \times \frac{2\ \cancel{L\,Cl_2}}{2\ \cancel{L\,Cl_2O_3}} \times \frac{1000\ mL\ Cl_2}{1\ \cancel{L\,Cl_2}} = 1750\ mL\ Cl_2$$

45. $2\,SO_2(g) + O_2(g) \rightarrow 2\,SO_3(g)$

$$25.0\ \cancel{L\,O_2} \times \frac{2\ L\ SO_3}{1\ \cancel{L\,O_2}} = 50.0\ L\ SO_3$$

47. $2\,N_2(g) + 5\,O_2(g) \rightarrow 2\,N_2O_5(g)$

$$500.0\ \cancel{cm^3\,N_2O_5} \times \frac{2\ cm^3\ N_2}{2\ \cancel{cm^3\,N_2O_5}} = 500.0\ cm^3\ N_2$$

Section 9.7 *Limiting Reactant Concept*

49. $N_2(g) + O_2(g) \rightarrow 2\,NO(g)$

$$5.00\ \cancel{mol\ N_2} \times \frac{2\ mol\ NO}{1\ \cancel{mol\ N_2}} = 10.0\ mol\ NO$$

$$5.00\ \cancel{mol\ O_2} \times \frac{2\ mol\ NO}{1\ \cancel{mol\ O_2}} = 10.0\ mol\ NO$$

N_2 or O_2 are each a limiting reactant because each produces the same amount of product. This reaction produces 10.0 mol NO.

51. $2\,NO(g) + O_2(g) \rightarrow 2\,NO_2(g)$

$$1.50\ \cancel{mol\ NO} \times \frac{2\ mol\ NO_2}{2\ \cancel{mol\ NO}} = 1.50\ mol\ NO_2$$

$$1.50\ \cancel{mol\ O_2} \times \frac{2\ mol\ NO_2}{1\ \cancel{mol\ O_2}} = 3.00\ mol\ NO_2$$

The limiting reactant is NO because it produces less product. This reaction produces 1.50 mol NO_2.

53. $2\,H_2(g) + O_2(g) \rightarrow 2\,H_2O(g)$

$$5.00\ \cancel{\text{mol } H_2} \times \frac{2\ \text{mol } H_2O}{2\ \cancel{\text{mol } H_2}} = 5.00\ \text{mol } H_2O$$

$$5.00\ \cancel{\text{mol } O_2} \times \frac{2\ \text{mol } H_2O}{1\ \cancel{\text{mol } O_2}} = 10.0\ \text{mol } H_2O$$

The limiting reactant is H_2 because it produces less product. This reaction produces 5.00 mol H_2O.

55. $2\,C_2H_6(g) + 7\,O_2(g) \rightarrow 4\,CO_2(g) + 6\,H_2O(g)$

$$1.00\ \cancel{\text{mol } C_2H_6} \times \frac{6\ \text{mol } H_2O}{2\ \cancel{\text{mol } C_2H_6}} = 3.00\ \text{mol } H_2O$$

$$5.00\ \cancel{\text{mol } O_2} \times \frac{6\ \text{mol } H_2O}{7\ \cancel{\text{mol } O_2}} = 4.29\ \text{mol } H_2O$$

The limiting reactant is C_2H_6 because it produces less water. This reaction produces 3.00 mol H_2O.

57. $Co(s) + S(s) \rightarrow CoS(s)$

(a) $$1.00\ \cancel{\text{mol Co}} \times \frac{1\ \text{mol CoS}}{1\ \cancel{\text{mol Co}}} = 1.00\ \text{mol CoS}$$

$$1.00\ \cancel{\text{mol S}} \times \frac{1\ \text{mol CoS}}{1\ \cancel{\text{mol S}}} = 1.00\ \text{mol CoS}$$

Neither Co or S is a limiting reactant because each produces the same amount of product. This reaction produces 1.00 mol CoS.

<u>After reaction:</u>

mol of Co	=	1.00 – 1.00	= 0.00 mol
mol of S	=	1.00 – 1.00	= 0.00 mol
mol of CoS	=	0.00 + 1.00	= 1.00 mol

(b) $$2.00\ \cancel{\text{mol Co}} \times \frac{1\ \text{mol CoS}}{1\ \cancel{\text{mol Co}}} = 2.00\ \text{mol CoS}$$

$$3.00\ \cancel{\text{mol S}} \times \frac{1\ \text{mol CoS}}{1\ \cancel{\text{mol S}}} = 3.00\ \text{mol CoS}$$

The limiting reactant is Co because it produces less product. This reaction produces 2.00 mol CoS.

<u>After reaction:</u>

mol of Co	=	2.00 – 2.00	=	0.00 mol
mol of S	=	3.00 – 2.00	=	1.00 mol
mol of CoS	=	0.00 + 2.00	=	2.00 mol

Section 9.8 *Limiting Reactant Problems*

59. $FeO(l) + Mg(l) \rightarrow Fe(l) + MgO(s)$

$$50.0\ \cancel{\text{g FeO}} \times \frac{1\ \cancel{\text{mol FeO}}}{71.85\ \cancel{\text{g FeO}}} \times \frac{1\ \cancel{\text{mol Fe}}}{1\ \cancel{\text{mol FeO}}} \times \frac{55.85\ \text{g Fe}}{1\ \cancel{\text{mol Fe}}} = 38.9\ \text{g Fe}$$

$$10.0\ \cancel{\text{g Mg}} \times \frac{1\ \cancel{\text{mol Mg}}}{24.31\ \cancel{\text{g Mg}}} \times \frac{1\ \cancel{\text{mol Fe}}}{1\ \cancel{\text{mol Mg}}} \times \frac{55.85\ \text{g Fe}}{1\ \cancel{\text{mol Fe}}} = 23.0\ \text{g Fe}$$

Therefore, Mg is the limiting reactant and will be consumed before FeO. This reaction produces 23.0 g Fe.

61. $Fe_2O_3(l) + 2\ Al(l) \rightarrow 2\ Fe(l) + Al_2O_3(g)$

$$50.0\ \cancel{\text{g Fe}_2\text{O}_3} \times \frac{1\ \cancel{\text{mol Fe}_2\text{O}_3}}{159.70\ \cancel{\text{g Fe}_2\text{O}_3}} \times \frac{2\ \cancel{\text{mol Fe}}}{1\ \cancel{\text{mol Fe}_2\text{O}_3}} \times \frac{55.85\ \text{g Fe}}{\cancel{\text{mol Fe}}} = 35.0\ \text{g Fe}$$

$$15.0\ \cancel{\text{g Al}} \times \frac{1\ \cancel{\text{mol Al}}}{26.98\ \cancel{\text{g Al}}} \times \frac{2\ \cancel{\text{mol Fe}}}{2\ \cancel{\text{mol Al}}} \times \frac{55.85\ \text{g Fe}}{1\ \cancel{\text{mol Fe}}} = 31.1\ \text{g Fe}$$

Therefore, Al is the limiting reactant because it will be consumed before Fe_2O_3. This reaction produces 31.1 g Fe.

63. $Mg(OH)_2(s) + H_2SO_4(l) \rightarrow MgSO_4(s) + 2\ H_2O(l)$

$$1.00\ \cancel{\text{g Mg(OH)}_2} \times \frac{1\ \cancel{\text{mol Mg(OH)}_2}}{58.33\ \cancel{\text{g Mg(OH)}_2}} \times \frac{1\ \cancel{\text{mol MgSO}_4}}{1\ \cancel{\text{mol Mg(OH)}_2}} \times \frac{120.38\ \text{g MgSO}_4}{1\ \cancel{\text{mol MgSO}_4}}$$

$$= 2.06\ \text{g MgSO}_4$$

$$0.605\ \cancel{\text{g H}_2\text{SO}_4} \times \frac{1\ \cancel{\text{mol H}_2\text{SO}_4}}{98.09\ \cancel{\text{g H}_2\text{SO}_4}} \times \frac{1\ \cancel{\text{mol MgSO}_4}}{1\ \cancel{\text{mol H}_2\text{SO}_4}} \times \frac{120.38\ \text{g MgSO}_4}{1\ \cancel{\text{mol MgSO}_4}}$$

$$= 0.742\ \text{g MgSO}_4$$

Therefore, H_2SO_4 is the limiting reactant because it will be consumed before $Mg(OH)_2$. This reaction produces 0.742 g $MgSO_4$.

65. $2\ Al(OH)_3(s) + 3\ H_2SO_4(l) \rightarrow Al_2(SO_4)_3(aq) + 6\ H_2O(l)$

$$1.00\ \cancel{g\ Al(OH)_3} \times \frac{1\ \cancel{mol\ Al(OH)_3}}{78.01\ \cancel{g\ Al(OH)_3}} \times \frac{6\ \cancel{mol\ H_2O}}{2\ \cancel{mol\ Al(OH)_3}} \times \frac{18.02\ g\ H_2O}{1\ \cancel{mol\ H_2O}} = 0.693\ g\ H_2O$$

$$3.00\ \cancel{g\ H_2SO_4} \times \frac{1\ \cancel{mol\ H_2SO_4}}{98.09\ \cancel{g\ H_2SO_4}} \times \frac{6\ \cancel{mol\ H_2O}}{3\ \cancel{mol\ H_2SO_4}} \times \frac{18.02\ g\ H_2O}{1\ \cancel{mol\ H_2O}} = 1.10\ g\ H_2O$$

Therefore, $Al(OH)_3$ is the limiting reactant because it will be consumed before H_2SO_4. This reaction produces 0.693 g H_2O.

67. $N_2(g) + 2\ O_2(g) \rightarrow 2\ NO_2(g)$

$$45.0\ \cancel{mL\ N_2} \times \frac{2\ mL\ NO_2}{1\ \cancel{mL\ N_2}} = 90.0\ mL\ NO_2$$

$$95.0\ \cancel{mL\ O_2} \times \frac{2\ mL\ NO_2}{2\ \cancel{mL\ O_2}} = 95.0\ mL\ NO_2$$

Therefore, N_2 is the limiting reactant because it will be consumed before O_2. This reaction produces 90.0 mL NO_2.

69. $2\ N_2(g) + 3\ O_2(g) \rightarrow 2\ N_2O_3(g)$

$$70.0\ \cancel{mL\ N_2} \times \frac{2\ mL\ N_2O_3}{2\ \cancel{mL\ N_2}} = 70.0\ mL\ N_2O_3$$

$$45.0\ \cancel{mL\ O_2} \times \frac{2\ mL\ N_2O_3}{3\ \cancel{mL\ O_2}} = 30.0\ mL\ N_2O_3$$

Therefore, O_2 is the limiting reactant because it will be consumed before N_2. This reaction produces 30.0 mL N_2O_3.

71. $2\ SO_2(g) + O_2(g) \rightarrow 2\ SO_3(g)$

$$3.00\ \cancel{L\ SO_2} \times \frac{2\ L\ SO_3}{2\ \cancel{L\ SO_2}} = 3.00\ L\ SO_3$$

$$1.25\ \cancel{L\ O_2} \times \frac{2\ L\ SO_3}{1\ \cancel{L\ O_2}} = 2.50\ L\ SO_3$$

Therefore, O_2 is the limiting reactant because it will be consumed before SO_2. This reaction produces 2.50 L SO_3.

73. $4\,HCl(g) + O_2(g) \rightarrow 2\,Cl_2(g) + 2\,H_2O(g)$

$$50.0\ \text{L HCl} \times \frac{2\ \text{L Cl}_2}{4\ \text{L HCl}} = 25.0\ \text{L Cl}_2$$

$$10.0\ \text{L O}_2 \times \frac{2\ \text{L Cl}_2}{1\ \text{L O}_2} = 20.0\ \text{L Cl}_2$$

Therefore, O_2 is the limiting reactant because it will be consumed before HCl. This reaction produces 20.0 L Cl_2.

Section 9.9 *Percent Yield*

75. Actual yield: 0.988 g Na_2CO_3; theoretical yield: 0.946 g Na_2CO_3

$$\text{Percent yield: } \frac{0.988\ \text{g}}{0.946\ \text{g}} \times 100\% = 104\%$$

77. Actual yield: 1.29 g $NaNO_2$; theoretical yield: 1.22 g $NaNO_2$

$$\text{Percent yield: } \frac{1.29\ \text{g}}{1.22\ \text{g}} \times 100\% = 106\%$$

General Exercises

79. The units of molar mass are grams per mole (g/mol).

81. $(NH_4)_2Cr_2O_7(s) \rightarrow Cr_2O_3(s) + 4\,H_2O(l) + N_2(g)$

MM of $(NH_4)_2Cr_2O_7$ = 252.10 g/mol
MM of Cr_2O_3 = 152.00 g/mol

$$1.54\ \text{g (NH}_4)_2\text{Cr}_2\text{O}_7 \times \frac{1\ \text{mol (NH}_4)_2\text{Cr}_2\text{O}_7}{252.10\ \text{g (NH}_4)_2\text{Cr}_2\text{O}_7} \times \frac{1\ \text{mol Cr}_2\text{O}_3}{1\ \text{mol (NH}_4)_2\text{Cr}_2\text{O}_7}$$

$$\times \frac{152.00\ \text{g Cr}_2\text{O}_3}{1\ \text{mol Cr}_2\text{O}_3} = 0.929\ \text{g Cr}_2\text{O}_3$$

83. $3\ MnO_2(l) + 4\ Al(l) \rightarrow 3\ Mn(l) + 2\ Al_2O_3(s)$

MM of Mn = 54.94 g/mol
MM of Al = 26.98 g/mol

$$1.00\ \cancel{kg\ Mn} \times \frac{1000\ \cancel{g\ Mn}}{1\ \cancel{kg\ Mn}} \times \frac{1\ \cancel{mol\ Mn}}{54.94\ \cancel{g\ Mn}} \times \frac{4\ \cancel{mol\ Al}}{3\ \cancel{mol\ Mn}} \times \frac{26.98\ g\ Al}{1\ \cancel{mol\ Al}}$$

$$= 655\ g\ Al$$

85. $Sb_2S_3(s) + 6\ HCl(aq) \rightarrow 2\ SbCl_3(aq) + 3\ H_2S(g)$

MM of Sb_2S_3 = 339.71 g/mol

$$3.00\ \cancel{g\ Sb_2S_3} \times \frac{1\ \cancel{mol\ Sb_2S_3}}{339.71\ \cancel{g\ Sb_2S_3}} \times \frac{3\ \cancel{mol\ H_2S}}{1\ \cancel{mol\ Sb_2S_3}} \times \frac{22.4\ L\ H_2S}{1\ \cancel{mol\ H_2S}} = 0.593\ L\ H_2S$$

Challenge Exercises

87. $C_3H_8(g) + 5\ O_2(g) \rightarrow 3\ CO_2(g) + 4\ H_2O(g)$

MM of C_3H_8 = 44.11 g/mol
MM of H_2O = 18.02 g/mol

$$10.0\ \cancel{g\ C_3H_8} \times \frac{1\ \cancel{mol\ C_3H_8}}{44.11\ \cancel{g\ C_3H_8}} \times \frac{4\ \cancel{mol\ H_2O}}{1\ \cancel{mol\ C_3H_8}} \times \frac{18.02\ g\ H_2O}{1\ \cancel{mol\ H_2O}} = 16.3\ g\ H_2O$$

89. $2\ H_2O(l) \rightarrow 2\ H_2(g) + O_2(g)$

MM of H_2O = 18.02 g/mol

$$100.0\ \cancel{mL\ H_2O} \times \frac{1.00\ \cancel{g\ H_2O}}{1\ \cancel{mL\ H_2O}} \times \frac{1\ \cancel{mol\ H_2O}}{18.02\ \cancel{g\ H_2O}} \times \frac{2\ \cancel{mol\ H_2}}{2\ \cancel{mol\ H_2O}} \times \frac{22.4\ L\ H_2}{1\ \cancel{mol\ H_2}}$$

$$= 124\ L\ H_2$$

CHAPTER 10

Gases

Section 10.1 *Properties of Gases*

1. The following are observed properties of gases:
   (a) Gases have a variable shape.
   (b) Gases expand uniformly.
   (c) Gases do **not** compress infinitely; they condense to a liquid.
   (d) Gases mix with other gases.

Section 10.2 *Atmospheric Pressure*

3.

	Units	Standard Pressure
(a)	millimeters of mercury	760 mm Hg
(b)	pounds per square inch	14.7 psi

5. (a) $2.50\ \cancel{\text{atm}} \times \frac{760\ \text{mm Hg}}{1\ \cancel{\text{atm}}} = 1.90 \times 10^3\ \text{mm Hg}$

   (b) $2.50\ \cancel{\text{atm}} \times \frac{14.7\ \text{psi}}{1\ \cancel{\text{atm}}} = 36.8\ \text{psi}$

7. (a) $30.8\ \cancel{\text{in. Hg}} \times \frac{101\ \text{kPa}}{29.9\ \cancel{\text{in. Hg}}} = 104\ \text{kPa}$

   (b) $30.8\ \cancel{\text{in. Hg}} \times \frac{760\ \text{torr}}{29.9\ \cancel{\text{in. Hg}}} = 783\ \text{torr}$

Section 10.3 *Variables Affecting Gas Pressure*

9.

	Change	Observation	Explanation
(a)	volume increases	pressure decreases	molecules collide less frequently
(b)	temperature increases	pressure increases	molecules collide more frequently
(c)	moles of gas increase	pressure increases	more molecules have more collisions

11.

	Change	Observation
(a)	volume increases	pressure decreases
(b)	temperature increases	pressure increases
(c)	moles of gas increases	pressure increases

13. Increasing the volume of a gas causes gas molecules to be farther apart, and they collide with the container less frequently. Since there are fewer collisions, the pressure of the gas decreases.

Section 10.4 *Boyle's Law: Pressure–Volume Relationships*

15. Pressure vs. Volume

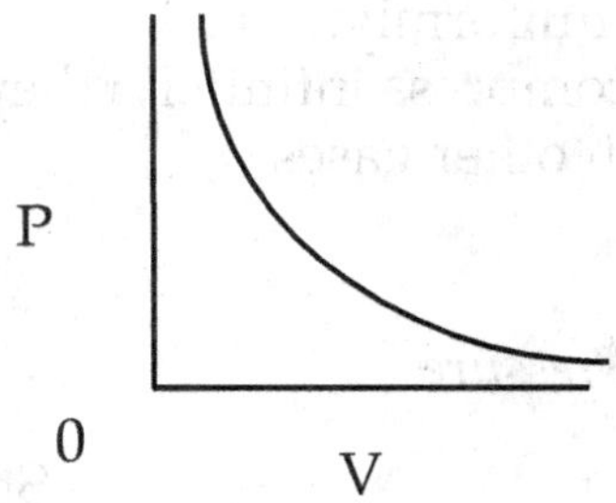

17. $P_1 \times V_{\text{factor}} = P_2$

$$5.00 \text{ atm} \times \frac{1.75 \cancel{\text{L}}}{2.50 \cancel{\text{L}}} = 3.50 \text{ atm}$$

19. $V_1 \times P_{\text{factor}} = V_2$

$$5.00 \text{ L} \times \frac{1.55 \cancel{\text{atm}}}{6.50 \cancel{\text{atm}}} = 1.19 \text{ L}$$

Section 10.5 *Charles's Law: Volume–Temperature Relationships*

21. Volume vs. Kelvin Temperature

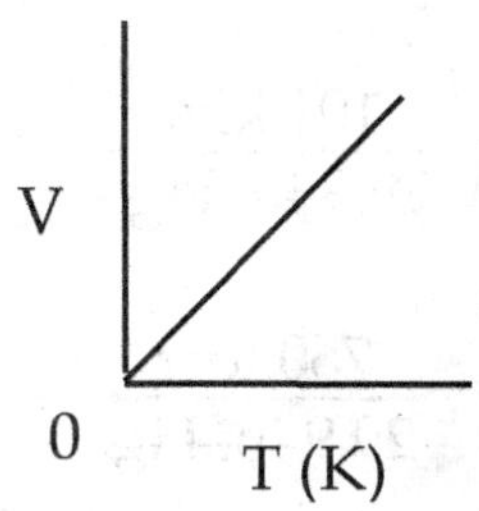

23. $V_1 \times T_{\text{factor}} = V_2$

25 °C + 273 = 298 K
50 °C + 273 = 323 K

$$555\ \text{mL} \times \frac{323\ \cancel{\text{K}}}{298\ \cancel{\text{K}}} = 602\ \text{mL}$$

25. $T_1 \times V_{\text{factor}} = T_2$

50 °C + 273 = 323 K

$$323\ \text{K} \times \frac{3.75\ \cancel{\text{L}}}{5.00\ \cancel{\text{L}}} = 242\ \text{K} \qquad 242\ \text{K} - 273 = -31\ °\text{C}$$

Section 10.6 *Gay-Lussac's Law: Pressure–Temperature Relationships*

27. <u>Pressure vs. Kelvin Temperature</u>

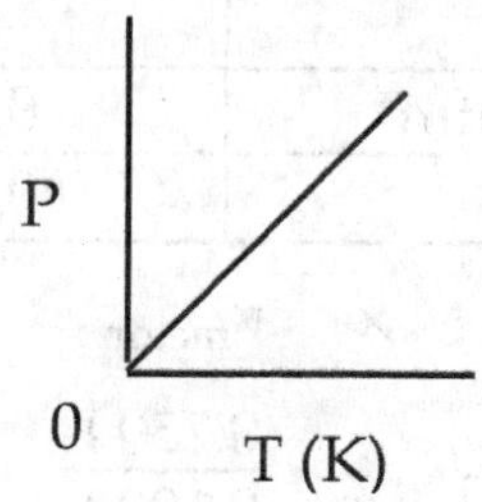

29. $P_1 \times T_{\text{factor}} = P_2$

20 °C + 273 = 293 K
110 °C + 273 = 383 K

$$1.00\ \text{atm} \times \frac{383\ \cancel{\text{K}}}{293\ \cancel{\text{K}}} = 1.31\ \text{atm}$$

31. $T_1 \times P_{\text{factor}} = T_2$

20 °C + 273 = 293 K

$$293\ \text{K} \times \frac{0.750\ \cancel{\text{atm}}}{0.500\ \cancel{\text{atm}}} = 440\ \text{K} \qquad 440\ \text{K} - 273 = 167\ °\text{C}$$

Section 10.7 *Combined Gas Law*

33.

	P	V	T
initial	5.00 atm	5.00 L	500 °C + 273 = 773 K
final	1.00 atm	V_2	273 K

$$V_1 \times P_{factor} \times T_{factor} = V_2$$

$$5.00\text{ L} \times \frac{5.00\ \cancel{\text{atm}}}{1.00\ \cancel{\text{atm}}} \times \frac{273\ \cancel{\text{K}}}{773\ \cancel{\text{K}}} = 8.83\text{ L}$$

35.

	P	V	T
initial	155 mm Hg	2.00 L	-50 °C + 273 = 223 K
final	365 mm Hg	V_2	75 °C + 273 = 348 K

$$V_1 \times P_{factor} \times T_{factor} = V_2$$

$$2.00\text{ L} \times \frac{155\ \cancel{\text{mm Hg}}}{365\ \cancel{\text{mm Hg}}} \times \frac{348\ \cancel{\text{K}}}{223\ \cancel{\text{K}}} = 1.33\text{ L}$$

37.

	P	V	T
initial	1.00 atm	0.750 L	273 K
final	P_2	0.100 L	25 °C + 273 = 298 K

$$P_1 \times V_{factor} \times T_{factor} = P_2$$

$$1.00\text{ atm} \times \frac{0.750\ \cancel{\text{L}}}{0.100\ \cancel{\text{L}}} \times \frac{298\ \cancel{\text{K}}}{273\ \cancel{\text{K}}} = 8.19\text{ atm}$$

39.

	P	V	T
initial	1.00 atm	1.00 L	273 K
final	2.00 atm	10.0 L	T_2

$$T_1 \times P_{factor} \times V_{factor} = T_2$$

$$273\text{ K} \times \frac{2.00\ \cancel{\text{atm}}}{1.00\ \cancel{\text{atm}}} \times \frac{10.0\ \cancel{\text{L}}}{1.00\ \cancel{\text{L}}} = 5460\text{ K (5190 °C)}$$

41.

	P	V	T
initial	1.00 atm	2.75 L	273 K
final	P_2	4.55 L	45 °C + 273 = 318 K

$$P_1 \times V_{factor} \times T_{factor} = P_2$$

$$1.00\text{ atm} \times \frac{2.75\ \cancel{\text{L}}}{4.55\ \cancel{\text{L}}} \times \frac{318\ \cancel{\text{K}}}{273\ \cancel{\text{K}}} = 0.704\text{ atm}$$

Section 10.8 *The Vapor Pressure Concept*

43. The vapor pressure of water is much greater than the vapor pressure of mercury at 25 °C. Thus, the vapor pressure of water can be measured by placing a drop of water on a column of mercury in a barometer.

45.

	Temperature	Vapor Pressure
(a)	25 °C	23.8 mm Hg
(b)	50 °C	92.5 mm Hg

Section 10.9 *Dalton's Law of Partial Pressures*

47.
$P_{\text{nitrogen}} = 592\text{ mm Hg}$
$P_{\text{oxygen}} = 160\text{ mm Hg}$
$P_{\text{argon}} = 7\text{ mm Hg}$
$P_{\text{trace gas}} = 1\text{ mm Hg}$

$$P_{\text{nitrogen}} + P_{\text{oxygen}} + P_{\text{argon}} + P_{\text{trace gas}} = P_{\text{atmosphere}}$$
$$592\text{ mm Hg} + 160\text{ mm Hg} + 7\text{ mm Hg} + 1\text{ mm Hg} = 760\text{ mm Hg}$$

49.
$P_{\text{total}} = 1\text{ atm} = 760\text{ mm Hg}$
$P_{\text{sulfur dioxide}} = 150\text{ mm Hg}$
$P_{\text{sulfur trioxide}} = 475\text{ mm Hg}$

$$P_{\text{sulfur dioxide}} + P_{\text{sulfur trioxide}} + P_{\text{oxygen}} = P_{\text{total}}$$
$$P_{\text{oxygen}} = P_{\text{total}} - P_{\text{sulfur dioxide}} - P_{\text{sulfur trioxide}}$$
$$P_{\text{oxygen}} = 760\text{ mm Hg} - 150\text{ mm Hg} - 475\text{ mm Hg} = 135\text{ mm Hg}$$

51. A gas may be collected over water to determine its volume. The volume of gas collected displaces an equal volume of water (see textbook Figure 10.11). When collecting a gas over water to determine volume, we must assume that the gas is not very soluble in water.

53.
$P_{\text{total}} = 766\text{ mm Hg}$
$P_{\text{water vapor}} = 17.5\text{ mm Hg at }20\text{ °C}$

$$P_{\text{water vapor}} + P_{\text{oxygen}} = P_{\text{total}}$$
$$P_{\text{oxygen}} = P_{\text{total}} - P_{\text{water vapor}}$$
$$P_{\text{oxygen}} = 766\text{ mm Hg} - 17.5\text{ mm Hg} = 749\text{ mm Hg}$$

Section 10.10 *Ideal Gas Behavior*

55. An ideal gas at absolute zero (0 K) exerts zero pressure (0 mm Hg).

57. An ideal gas at absolute zero (0 K) possesses zero kinetic energy.

59. According to the kinetic theory, the characteristics of an ideal gas are as follows:
    (1) An ideal gas is mostly empty space and composed of tiny molecules, which have a negligible volume.
    (2) An ideal gas has molecules that demonstrate rapid motion, move in straight lines, and travel in random directions.
    (3) An ideal gas has molecules that show no attraction for one another.
    (4) An ideal gas has molecules that demonstrate elastic collisions.
    (5) In an ideal gas the average kinetic energy of molecules is proportional to the Kelvin temperature.

61.

	Description	Gas
(a)	highest kinetic energy	all gases are equal
(b)	lowest kinetic energy	all gases are equal
(c)	fastest atoms	He atoms (lightest atoms)
(d)	slowest atoms	Ar atoms (heaviest atoms)

Section 10.11 *Ideal Gas Law*

63. $n = \dfrac{P\,V}{R\,T}$

$V = 5.00\ \text{L}$

$T = 373\ \text{K}$

$P = 1.75\ \text{atm}$

$$n = \frac{1.75\ \cancel{\text{atm}} \times 5.00\ \cancel{\text{L}}}{373\ \cancel{\text{K}}} \times \frac{1\ \text{mol} \cdot \cancel{\text{K}}}{0.0821\ \cancel{\text{atm}} \cdot \cancel{\text{L}}} = 0.286\ \text{mol He}$$

65. $P = \dfrac{n\,R\,T}{V}$

$n = 0.500\ \text{mol He}$

$T = 25\ °\text{C} + 273 = 298\ \text{K}$

$V = 5.00\ \text{L}$

$$P = \frac{0.500\ \cancel{\text{mol}} \times 298\ \cancel{\text{K}}}{5.00\ \cancel{\text{L}}} \times \frac{0.0821\ \text{atm} \cdot \cancel{\text{L}}}{1\ \cancel{\text{mol}} \cdot \cancel{\text{K}}} = 0.411\ \text{atm}$$

General Exercises

67. $$76.0\ \cancel{\text{cm Hg}} \times \frac{13.6\ \text{cm H}_2\text{O}}{1\ \cancel{\text{cm Hg}}} = 1030\ \text{cm H}_2\text{O}$$

69. $$P_{\text{total}} = 35\ \text{atm} + 125\ \text{atm} = 160\ \text{atm}$$

$$P_{\text{total}} = 160\ \cancel{\text{atm}} \times \frac{14.7\ \text{psi}}{1\ \cancel{\text{atm}}} = 2350\ \text{psi}$$

71. $$P_{\text{carbon dioxide}} + P_{\text{water vapor}} = P_{\text{total}}$$

$$P_{\text{oxygen}} = P_{\text{total}} - P_{\text{water vapor}}$$

$$P_{\text{oxygen}} = 764\ \text{mm Hg} - 19.8\ \text{mm Hg} = 744\ \text{mm Hg}$$

	P	V	T
initial	744 mm Hg	42.5 mL	22 °C + 273 = 295 K
final	760 mm Hg	V_2	273 K

$$42.5\ \text{mL} \times \frac{744\ \cancel{\text{mm Hg}}}{760\ \cancel{\text{mm Hg}}} \times \frac{273\ \cancel{\text{K}}}{295\ \cancel{\text{K}}} = 38.5\ \text{mL}$$

73. $$P = \frac{n\,R\,T}{V}$$

$$n = 1.51 \times 10^{23}\ \cancel{\text{molecules}} \times \frac{1\ \text{mol}}{6.02 \times 10^{23}\ \cancel{\text{molecules}}} = 0.251\ \text{mol}$$

$T = 25\ °\text{C} + 273 = 298\ \text{K}$ $\qquad V = 5.00\ \text{L}$

$$P = \frac{0.251\ \cancel{\text{mol}} \times 298\ \cancel{\text{K}}}{5.00\ \cancel{\text{L}}} \times \frac{0.0821\ \text{atm} \cdot \cancel{\text{L}}}{1\ \cancel{\text{mol}} \cdot \cancel{\text{K}}} = 1.23\ \text{atm}$$

75. Methane, CH_4, has a lower molecular mass than both ethane, C_2H_6, and hydrogen sulfide, H_2S; therefore, CH_4 molecules have the fastest velocity.

77. $$n = \frac{P\,V}{R\,T} \qquad V = 1550\ \cancel{\text{mL}} \times \frac{1\ \text{L}}{1000\ \cancel{\text{mL}}} = 1.55\ \text{L}$$

$$P = 0.945\ \text{atm}$$

$$T = 50\ °\text{C} + 273 = 323\ \text{K}$$

$$n = \frac{0.945\ \cancel{\text{atm}} \times 1.55\ \cancel{\text{L}}}{323\ \cancel{\text{K}}} \times \frac{1\ \text{mol} \cdot \cancel{\text{K}}}{0.0821\ \cancel{\text{atm}} \cdot \cancel{\text{L}}} = 0.0552\ \text{mol}$$

$$0.0552\ \cancel{\text{mol Cl}_2} \times \frac{70.90\ \text{g Cl}_2}{1\ \cancel{\text{mol Cl}_2}} = 3.92\ \text{g Cl}_2$$

Challenge Exercises

79. $255 \cancel{\text{in.}^2} \times \dfrac{14.7 \text{ lb}}{1 \cancel{\text{in.}^2}} = 3750 \text{ lb}$

81. $MM = \dfrac{g\,R\,T}{P\,V}$

$g = 2.14 \text{ g}$

$V = 1.00 \text{ L}$

$P = 1 \text{ atm}$

$T = 273 \text{ K}$

$$\text{MM} = \frac{2.14 \text{ g} \times 273 \cancel{\text{K}}}{1 \cancel{\text{atm}} \times 1.00 \cancel{\text{L}}} \times \frac{0.0821 \cancel{\text{atm}} \cdot \cancel{\text{L}}}{1 \text{ mol} \cdot \cancel{\text{K}}} = 48.0 \text{ g/mol}$$

The molar mass of $\mathbf{O_3}$ is closest to the calculated value of 48.0 g/mol.

83. $MM = \dfrac{g\,R\,T}{P\,V}$

$g = 1.95 \text{ g}$

$V = 3.00 \text{ L}$

$P = 1.25 \text{ atm}$

$T = 20\text{ °C} + 273 = 293 \text{ K}$

$$MM = \frac{1.95 \text{ g} \times 293 \cancel{\text{K}}}{1.25 \cancel{\text{atm}} \times 3.00 \cancel{\text{L}}} \times \frac{0.0821 \cancel{\text{atm}} \cdot \cancel{\text{L}}}{1 \text{ mol} \cdot \cancel{\text{K}}} = 12.5 \text{ g/mol}$$

CHAPTER

11

Liquids and Solids

Section 11.1 *Properties of Liquids*

1. General Observed Properties of Liquids
   - (a) *True,* liquids have a fixed volume.
   - (b) *True,* liquids that are soluble mix homogeneously with each other.
   - (c) *False,* liquids do not compress significantly with changes in temperature or pressure; however, they may undergo a change in physical state.
   - (d) *True,* liquids are about 1000 times *more dense* than gases.

3.

	Substance	Temperature	Physical State
(a)	H_2O	–10 °C	solid
(b)	H_2O	110 °C	gas
(c)	NH_3	–100 °C	solid
(d)	NH_3	–50 °C	liquid

Section 11.2 *The Intermolecular Bond Concept*

5. A *temporary dipole* is caused by electrons moving about in a molecule. The resulting positive and negative charges in a molecule give rise to a temporary attraction between molecules. A *permanent dipole* is caused by electrons being drawn closer to one atom than another within a molecule. The resulting positive charge in one molecule is attracted to the negative change in another molecule.

Section 11.3 *Vapor Pressure, Boiling Point, Viscosity, and Surface Tension*

7. Water molecules in the liquid state evaporate and form water vapor. Simultaneously, water molecules in the vapor condense to a liquid. By definition, the vapor pressure of water is the total pressure exerted by the water molecules in the vapor when the *rate of evaporation* is equal to the *rate of condensation.*

9. The *viscosity* of a liquid measures its resistance to flow. Since the intermolecular attraction in water is strong, water has less tendency to flow than other liquids with molecules the same size.

11. If the molecules in a liquid are strongly attracted:
    (a) the vapor pressure is *low.* (b) the boiling point is *high.*
    (c) the viscosity is *high.* (d) the surface tension is *high.*

13.

	Liquid	Higher Vapor Pressure
(a)	CH_3COOH or C_2H_5Cl	C_2H_5Cl (weaker attraction)
(b)	C_2H_5OH or CH_3OCH_3	CH_3OCH_3 (weaker attraction)

15.

	Liquid	Higher Viscosity
(a)	CH_3COOH or C_2H_5Cl	CH_3COOH (stronger attraction)
(b)	C_2H_5OH or CH_3OCH_3	C_2H_5OH (stronger attraction)

17. The general relationship between the vapor pressure of a liquid and temperature is: *as the temperature of a liquid increases, the vapor pressure increases.*

19.

	Temperature	Vapor Pressure
(a)	45 °C	~180 mm Hg
(b)	60 °C	~330 mm Hg

21. The temperature at which the vapor pressure is equal to the atmospheric pressure is defined as the *boiling point.* Thus, the boiling point of acetone is 56 °C because the vapor pressure at this temperature is 760 mm Hg.

Section 11.4 *Properties of Solids*

23. General Observed Properties of Solids
    (a) *True,* solids have a fixed volume.
    (b) *True,* solids can have a noncrystalline structure.
    (c) *True,* solids do not expand significantly.
    (d) *False,* alloys mix in the liquid state and cool to a solid.

25.

	Substance	Temperature	Physical State
(a)	Ga	ice water, 0 °C	solid
(b)	Ga	boiling water, 100 °C	liquid

Section 11.5 *Crystalline Solids*

27. Three examples of crystalline solids are NaCl (ionic solid), Cl_2 (molecular solid), and Na (metallic solid).

29.

	Crystalline Solid	Type of Particles
(a)	ionic solid	ions
(b)	molecular solid	molecules
(c)	metallic solid	metal atoms

31.

	Crystalline Solid	Classification
(a)	S_8	molecular solid
(b)	SO_2	molecular solid
(c)	Cu	metallic solid
(d)	$AgNO_3$	ionic solid

Section 11.6 *Changes of Physical State*

33. Heating Curve for Ethanol

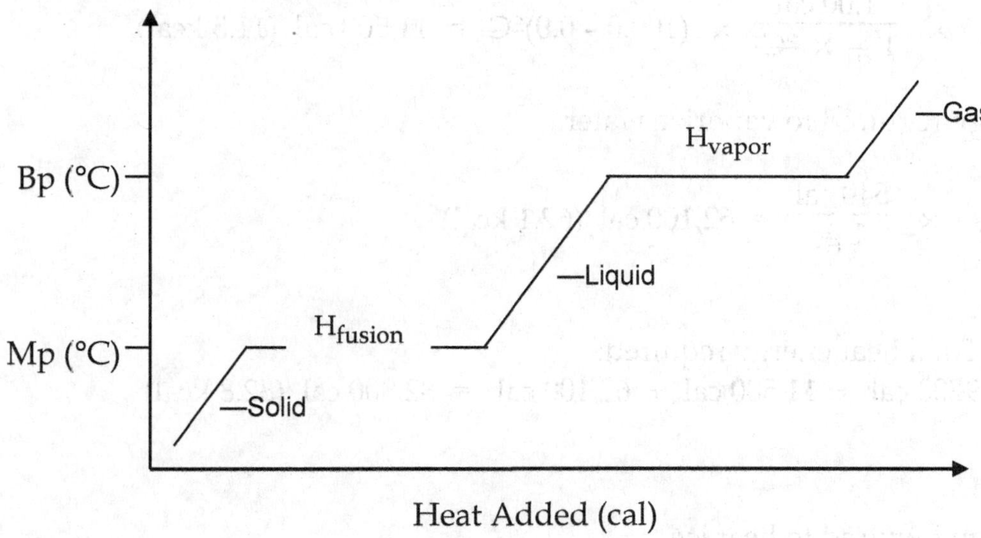

35. Energy *released* to change 25.0 g water to ice at 0 °C:

$$25.0\ \cancel{g} \times \frac{80.0\ \text{cal}}{1\ \cancel{g}} = 2.00 \times 10^3\ \text{cal}\ \ (2.00\ \text{kcal})$$

37. Energy required to heat water:

$$25.0\ \cancel{g} \times \frac{1.00\ cal}{1\ \cancel{g} \times \cancel{^\circ C}} \times (100.0 - 25.0)\cancel{^\circ C} = 1880\ cal\ \ (1.88\ kcal)$$

Energy required to vaporize water:

$$25.0\ \cancel{g} \times \frac{540\ cal}{1\ \cancel{g}} = 13{,}500\ cal\ \ (13.5\ kcal)$$

Total heat energy required:
1880 cal + 13,500 cal = 15,400 cal (15.4 kcal)

39. Energy required to melt ice:

$$115\ \cancel{g} \times \frac{80.0\ cal}{1\ \cancel{g}} = 9200\ cal\ \ (9.20\ kcal)$$

Energy required to heat water:

$$115\ \cancel{g} \times \frac{1.00\ cal}{1\ \cancel{g} \times \cancel{^\circ C}} \times (100.0 - 0.0)\cancel{^\circ C} = 11{,}500\ cal\ \ (11.5\ kcal)$$

Energy required to vaporize water:

$$115\ \cancel{g} \times \frac{540\ cal}{1\ \cancel{g}} = 62{,}100\ cal\ \ (62.1\ kcal)$$

Total heat energy required:
9200 cal + 11,500 cal + 62,100 cal = 82,800 cal (82.8 kcal)

41. Energy required to heat ice:

$$38.5\ \cancel{g} \times \frac{0.50\ cal}{1\ \cancel{g} \times \cancel{^\circ C}} \times [0.0 - (-20.0)]\cancel{^\circ C} = 385\ cal\ \ (0.385\ kcal)$$

Energy required to melt ice:

$$38.5\ \cancel{g} \times \frac{80.0\ cal}{1\ \cancel{g}} = 3080\ cal\ \ (3.08\ kcal)$$

Energy required to heat water:

$$38.5\,\cancel{g} \times \frac{1.00\text{ cal}}{1\,\cancel{g} \times \cancel{^{\circ}C}} \times (100.0 - 0.0)\cancel{^{\circ}C} = 3850\text{ cal } (3.85\text{ kcal})$$

Energy required to vaporize water:

$$38.5\,\cancel{g} \times \frac{540\text{ cal}}{1\,\cancel{g}} = 20{,}800\text{ cal } (20.8\text{ kcal})$$

Total heat energy required:
385 cal + 3080 cal + 3850 cal + 20,800 cal = 28,100 cal (28.1 kcal)

Section 11.7 *Structure of Water*

43. Water has an unusually *low* vapor pressure.

45. Water has an unusually *high* viscosity.

47. The observed bond angle in a water molecule is 104.5°.

49. Two Dipoles in a Water Molecule shown using delta notation

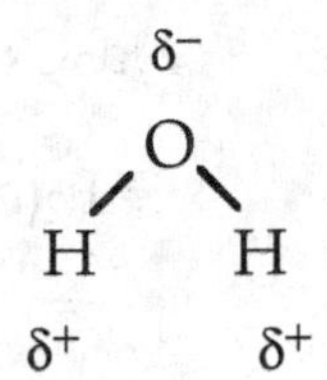

Section 11.8 *Physical Properties of Water*

51. *Water* has the highest specific heat of any liquid.

53. Since ice floats in water, *water* is more dense than ice.

55.

	Liquid	Higher Melting Point
(a)	H_2O or H_2S	H_2O (hydrogen bond attraction)
(b)	H_2S or H_2Se	H_2Se (larger size, more attraction)

57.

	Liquid	Higher Heat of Fusion
(a)	H_2O or H_2S	H_2O (hydrogen bond attraction)
(b)	H_2S or H_2Se	H_2Se (larger size, more attraction)

59.

	Property	As Molar Mass Increases
(a)	melting point	increases
(b)	boiling point	increases
(c)	heat of fusion	increases
(d)	heat of vaporization	increases

Section 11.9 *Chemical Properties of Water*

61. $2\,H_2(g) + O_2(g) \rightarrow 2\,H_2O(l)$

63. (a) $2\,Li(s) + 2\,H_2O(l) \rightarrow 2\,LiOH(aq) + H_2(g)$
(b) $Na_2O(s) + H_2O(l) \rightarrow 2\,NaOH(aq)$
(c) $CO_2(g) + H_2O(l) \rightarrow H_2CO_3(aq)$

65. (a) $Mg(s) + H_2O(l) \rightarrow NR$
(b) $SrO(s) + H_2O(l) \rightarrow Sr(OH)_2(aq)$
(c) $N_2O_5(s) + H_2O(l) \rightarrow 2\,HNO_3(aq)$

67. (a) $2\,C_3H_6(g) + 9\,O_2(g) \xrightarrow{\text{spark}} 6\,CO_2(g) + 6\,H_2O(g)$
(b) $Na_2Cr_2O_7 \cdot 2H_2O(s) \xrightarrow{\Delta} Na_2Cr_2O_7(s) + 2\,H_2O(g)$
(c) $2\,HF(aq) + Ca(OH)_2(aq) \rightarrow CaF_2(aq) + 2\,HOH(l)$

69. (a) $2\,C_3H_7OH(l) + 9\,O_2(g) \xrightarrow{\text{spark}} 6\,CO_2(g) + 8\,H_2O(g)$
(b) $Ca(NO_3)_2 \cdot 4H_2O(s) \xrightarrow{\Delta} Ca(NO_3)_2(s) + 4\,H_2O(g)$
(c) $H_2CO_3(aq) + 2\,KOH(aq) \rightarrow K_2CO_3(aq) + 2\,HOH(l)$

Section 11.10 *Hydrates*

71.

	Chemical Formula	Systematic Name
(a)	$MgCl_2 \cdot 6H_2O$	magnesium chloride hexahydrate
(b)	$MnSO_4 \cdot 7H_2O$	manganese(II) sulfate heptahydrate
(c)	$Co(CN)_3 \cdot 3H_2O$	cobalt(III) cyanide trihydrate
(d)	$Cr(NO_3)_3 \cdot 9H_2O$	chromium(III) nitrate nonahydrate

73.

	Systematic Name	Chemical Formula
(a)	sodium acetate trihydrate	$NaC_2H_3O_2 \cdot 3H_2O$
(b)	calcium sulfate dihydrate	$CaSO_4 \cdot 2H_2O$
(c)	potassium chromate tetrahydrate	$K_2CrO_4 \cdot 4H_2O$
(d)	zinc sulfate heptahydrate	$ZnSO_4 \cdot 7H_2O$

75. (a) Percentage of water in $MnSO_4 \cdot H_2O$

MM of $MnSO_4$ = 54.94 g + 32.07 g + 4(16.00 g) = 151.01 g

$$\text{Percentage of water: } \frac{18.02\text{ g}}{151.01\text{ g} + 18.02\text{ g}} \times 100\% = 10.66\%\ H_2O$$

(b) Percentage of water in $Sr(NO_3)_2 \cdot 6H_2O$

MM of $Sr(NO_3)_2$ = 87.62 g + 2(14.01 g) + 6(16.00 g) = 211.64 g

$$\text{Percentage of water: } \frac{6(18.02\text{ g})}{211.64\text{ g} + 6(18.02\text{ g})} \times 100\% = 33.81\%\ H_2O$$

(c) Percentage of water in $Co(C_2H_3O_2)_2 \cdot 4H_2O$

MM of $Co(C_2H_3O_2)_2$ = 58.93 g + 4(12.01 g) + 6(1.01 g) + 4(16.00 g) = 177.03 g

$$\text{Percentage of water: } \frac{4(18.02\text{ g})}{177.03\text{ g} + 4(18.02\text{ g})} \times 100\% = 28.94\%\ H_2O$$

(d) Percentage of water in $Cr(NO_3)_3 \cdot 9H_2O$

MM of $Cr(NO_3)_3$ = 52.00 g + 3(14.01 g) + 9(16.00 g) = 238.03 g

$$\text{Percentage of water: } \frac{9(18.02\text{ g})}{238.03\text{ g} + 9(18.02\text{ g})} \times 100\% = 40.52\%\ H_2O$$

77. (a) $SrCl_2 \cdot XH_2O(s) \xrightarrow{\Delta} SrCl_2(s) + X\,H_2O(g)$

$$18.5\ \cancel{g\ H_2O} \times \frac{1\ mol\ H_2O}{18.02\ \cancel{g\ H_2O}} = 1.03\ mol\ H_2O$$

$$81.5\ \cancel{g\ SrCl_2} \times \frac{1\ mol\ SrCl_2}{158.52\ \cancel{g\ SrCl_2}} = 0.514\ mol\ SrCl_2$$

$$SrCl_2 \cdot \frac{1.03}{0.514}\ H_2O \qquad \frac{1.03}{0.514} = 2.00 \approx 2$$

Chemical Formula: $SrCl_2 \cdot 2H_2O$

(b) $Sr(NO_3)_2 \cdot XH_2O(s) \xrightarrow{\Delta} Sr(NO_3)_2(s) + XH_2O(g)$

$$33.8\ \cancel{g\ H_2O} \times \frac{1\ mol\ H_2O}{18.02\ \cancel{g\ H_2O}} = 1.88\ mol\ H_2O$$

$$66.2\ \cancel{g\ Sr(NO_3)_2} \times \frac{1\ mol\ Sr(NO_3)_2}{211.64\ \cancel{g\ Sr(NO_3)_2}} = 0.313\ mol\ Sr(NO_3)_2$$

$$Sr(NO_3)_2 \cdot \frac{1.88}{0.313}\ H_2O \qquad \frac{1.88}{0.313} = 6.01 \approx 6$$

Chemical Formula: $Sr(NO_3)_2 \cdot 6H_2O$

(c) $Ca(NO_3)_2 \cdot XH_2O(s) \xrightarrow{\Delta} Ca(NO_3)_2(s) + XH_2O(g)$

$$30.5\ \cancel{g\ H_2O} \times \frac{1\ mol\ H_2O}{18.02\ \cancel{g\ H_2O}} = 1.69\ mol\ H_2O$$

$$69.5\ \cancel{g\ Ca(NO_3)_2} \times \frac{1\ mol\ Ca(NO_3)_2}{164.10\ \cancel{g\ Ca(NO_3)_2}} = 0.424\ mol\ Ca(NO_3)_2$$

$$Ca(NO_3)_2 \cdot \frac{1.69}{0.424}\ H_2O \qquad \frac{1.69}{0.424} = 3.99 \approx 4$$

Chemical Formula: $Ca(NO_3)_2 \cdot 4H_2O$

(d) $Na_2B_4O_7 \cdot XH_2O(s) \xrightarrow{\Delta} Na_2B_4O_7(s) + X\,H_2O(g)$

$$30.9\ \cancel{g\ H_2O} \times \frac{1\ mol\ H_2O}{18.02\ \cancel{g\ H_2O}} = 1.71\ mol\ H_2O$$

$$69.1\ \cancel{g\ Na_2B_4O_7} \times \frac{1\ mol\ Na_2B_4O_7}{201.22\ \cancel{g\ Na_2B_4O_7}} = 0.343\ mol\ Na_2B_4O_7$$

$$Na_2B_4O_7 \cdot \frac{1.71}{0.343}\ H_2O \qquad \frac{1.71}{0.343} = 4.99 \approx 5$$

Chemical Formula: $Na_2B_4O_7 \cdot 5H_2O$

General Exercises

79. Water comprises about two-thirds (~65%) of the human body.

81.

	Liquid	Intermolecular Attraction
(a)	C_5H_{12}	dispersion force
(b)	$CH_3CH_2CH_2–Cl$	dipole force
(c)	$CH_3–OH$	hydrogen bond
(d)	$CH_3–O–CH_3$	dipole force

83. The boiling point of water at 0.500 atm (380 mm Hg) is approximately 80 °C.

85. Sulfuric acid, H_2SO_4, has O–H bonds, so it can hydrogen bond and molecules show strong intermolecular attraction. Thus, sulfuric acid has high surface tension, minimal surface area, and sulfuric acid "raindrops" on Venus will be spherical.

Challenge Exercises

87. Mass of methyl alcohol:

$$5.00\ \cancel{L} \times \frac{1000\ \cancel{mL}}{1\ \cancel{L}} \times \frac{0.792\ g}{1\ \cancel{mL}} = 3{,}960\ g$$

Energy required to heat methyl alcohol:

$$3{,}960\ \cancel{g} \times \frac{0.610\ cal}{1\ \cancel{g} \cdot \cancel{^{\circ}C}} \times (65.0 - 25.0)\cancel{^{\circ}C} = 96{,}600\ cal\ (96.6\ kcal)$$

Energy required to vaporize methyl alcohol:

$$3{,}960\ \cancel{g} \times \frac{263\ cal}{1\ \cancel{g}} = 1{,}040{,}000\ cal\ (1{,}040\ kcal)$$

Total heat energy required:

$$96{,}600\ cal + 1{,}040{,}000\ cal = 1{,}140{,}000\ cal\ (1{,}140\ kcal)$$

CHAPTER 12

Chemical Bonding

Section 12.1 *The Chemical Bond Concept*

1.

Atom	Valence Electrons
Na	1 valence e^-
Cl	7 valence e^-

3.

Ion	Ionic Bond
Na^+	0 valence e^-
Cl^-	8 valence e^-

5.

	Compound	Bond		Compound	Bond
(a)	$AlCl_3$	ionic	(b)	H_2O	covalent
(c)	SO_3	covalent	(d)	$FeCO_3$	ionic

7.

	Substance	Particle		Substance	Particle
(a)	CH_3OH	molecule	(b)	$KClO_2$	formula unit
(c)	MnO_2	formula unit	(d)	$HC_2H_3O_2$	molecule

9.

	Substance	Particle		Substance	Particle
(a)	C_3H_8	molecule	(b)	Cr	atom
(c)	Fe_2O_3	formula unit	(d)	S_8	molecule

Section 12.2 *Ionic Bonds*

11.

	Ion	Ionic Charge		Ion	Ionic Charge
(a)	Li ion	1+ (Group 1)	(b)	Mg ion	2+ (Group 2)
(c)	Al ion	3+ (Group 13)	(d)	Sn ion	4+ (Group 14)

13.

	Ion	Ionic Charge		Ion	Ionic Charge
(a)	F ion	1– (Group 17)	(b)	Br ion	1– (Group 17)
(c)	S ion	2– (Group 16)	(d)	N ion	3– (Group 15)

15.

	Ion	Electron Configuration
(a)	Na^+	$1s^2\ 2s^2\ 2p^6$
(b)	Mg^{2+}	$1s^2\ 2s^2\ 2p^6$
(c)	Al^{3+}	$1s^2\ 2s^2\ 2p^6$
(d)	V^{4+}	$1s^2\ 2s^2\ 2p^6\ 3s^2\ 3p^6\ 3d^1$

17.

	Ion	Electron Configuration
(a)	F^-	$1s^2\ 2s^2\ 2p^6$
(b)	O^{2-}	$1s^2\ 2s^2\ 2p^6$
(c)	N^{3-}	$1s^2\ 2s^2\ 2p^6$
(d)	C^{4-}	$1s^2\ 2s^2\ 2p^6$

19.

	Ion	Isoelectronic Noble Gas
(a)	S^{2-}	Ar
(b)	Cl^-	Ar
(c)	K^+	Ar
(d)	Ca^{2+}	Ar

21.

	Ion	Isoelectronic Noble Gas
(a)	Li^+	He
(b)	Al^{3+}	Ne
(c)	Ca^{2+}	Ar
(d)	Mg^{2+}	Ne

23.

	Ion	Isoelectronic Noble Gas
(a)	Cl^-	Ar
(b)	I^-	Xe
(c)	S^{2-}	Ar
(d)	P^{3-}	Ar

[illegible]. (a) Li atomic radius > Li ion radius
(b) Mg atomic radius > Mg ion radius
(c) F atomic radius < F ion radius
(d) O atomic radius < O ion radius

27. The true statements are (b) and (c). The corrected statements are:
(a) Calcium ions and oxide ions bond by *ionic* electrostatic attraction.
(d) Breaking an ionic bond between calcium and oxygen *requires* energy.

Section 12.3 *Covalent Bonds*

29. (a) The bond length in H–Cl is less than the sum of the atomic radii.
(b) The bond length in O=O is less than the sum of the atomic radii.

31. The true statements are (a) and (d). The corrected statements are:
(b) Bonding electrons are distributed over the entire NO molecule.
(c) The bond length is *less than* the sum of the two atomic radii.

Section 12.4 *Electron Dot Formulas of Molecules*

33.

Molecule	Valence Electrons	Electron Dot	Structural Formula
(a) $\mathbf{H_2}$	1 + 1 = 2 e⁻	H : H	H—H
(b) $\mathbf{F_2}$	7 + 7 = 14 e⁻	$:\ddot{\underset{..}{F}}:\ddot{\underset{..}{F}}:$	F — F
(c) H**Br**	1 + 7 = 8 e⁻	$H:\ddot{\underset{..}{Br}}:$	H— Br
(d) $\mathbf{NH_3}$	5 + 3(1) = 8 e–	$H:\overset{H}{\underset{..}{\ddot{N}}}:H$	$H-\overset{H}{\overset{\vert}{N}}-H$

35.

Molecule	Valence Electrons	Electron Dot	Structural Formula
(a) N_2	$2(5) = 10\ e^-$	:N:::N:	N≡N
(b) PI_3	$5 + 3(7) = 26\ e^-$	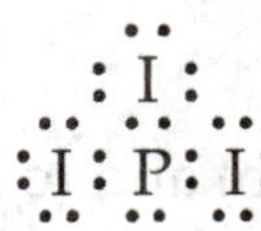	
(c) HONO	$1 + 2(6) + 5 = 18\ e^-$		H−O−N=O
(d) C_2H_4	$2(4) + 4(1) = 12\ e^-$		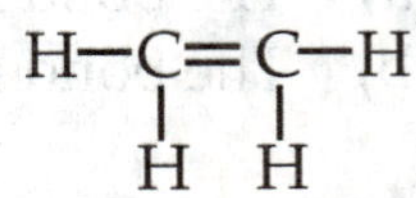

37.

Molecule	Valence Electrons	Electron Dot	Structural Formula
(a) CH_4	$4 + 4(1) = 8\ e^-$	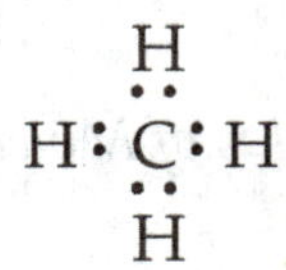	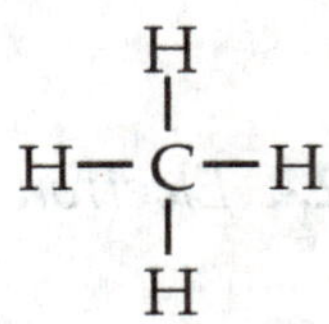
(b) OF_2	$6 + 2(7) = 20\ e^-$		F−O−F
(c) H_2O_2	$2(1) + 2(6) = 14\ e^-$		H−O−O−H
(d) NF_3	$5 + 3(7) = 26\ e^-$	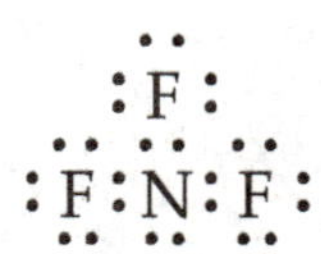	

ection 12.5 *Electron Dot Formulas of Polyatomic Ions*

39.

	Polyatomic Ion	Valence Electrons	Electron Dot	Structural Formula
(a)	IO^-	$7 + 6 + 1 = 14\ e^-$	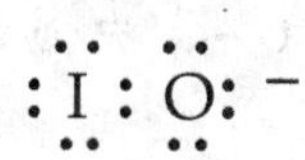	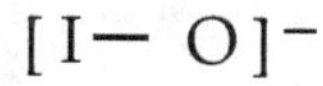
(b)	IO_2^-	$7 + 2(6) + 1 = 20\ e^-$		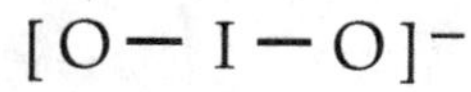
(c)	IO_3^-	$7 + 3(6) + 1 = 26\ e^-$	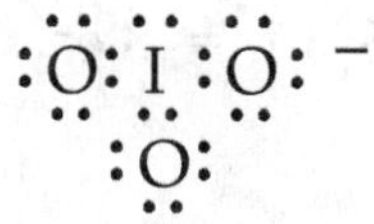	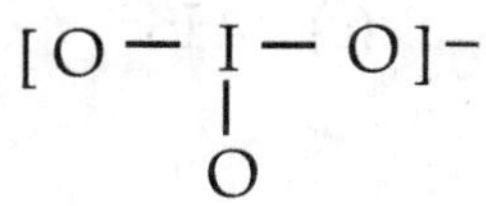
(d)	IO_4^-	$7 + 4(6) + 1 = 32\ e^-$	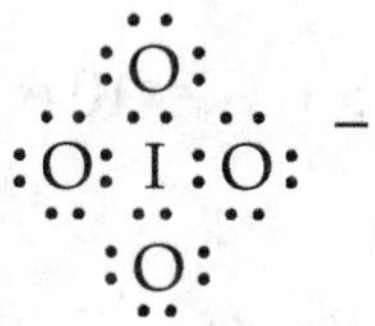	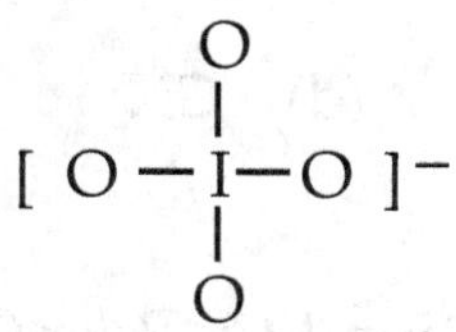

1.

	Polyatomic Ion	Valence Electrons	Electron Dot	Structural Formula
(a)	SO_4^{2-}	$6 + 4(6) + 2 = 32\ e^-$	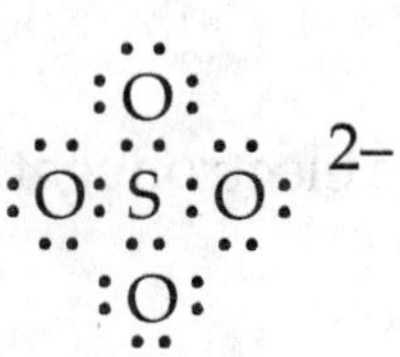	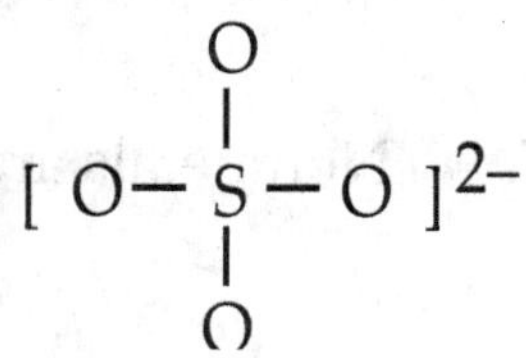
(b)	HSO_4^-	$1 + 6 + 4(6) + 1 = 32\ e^-$	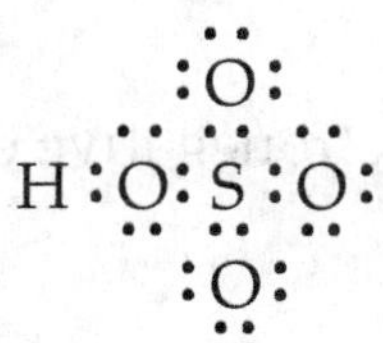	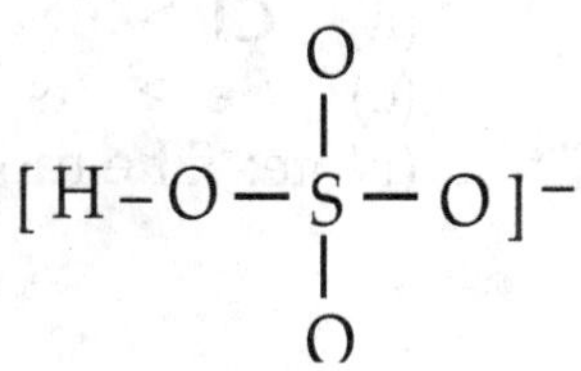
(c)	SO_3^{2-}	$6 + 3(6) + 2 = 26\ e^-$	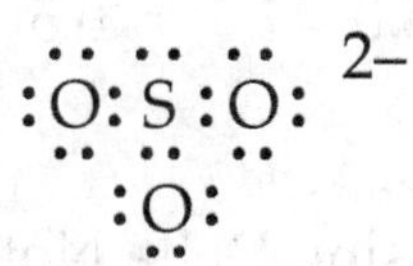	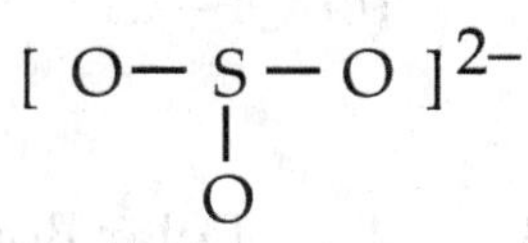
(d)	HSO_3^-	$1 + 6 + 3(6) + 1 = 26\ e^-$	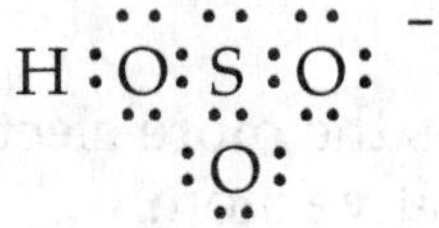	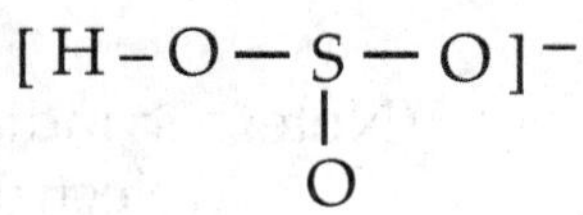

43.

	Polyatomic Ion	Valence Electrons	Electron Dot	Structural Formula
(a)	H_3O^+	$3(1) + 6 - 1 = 8\ e^-$	$\left[H:\underset{\cdot\cdot}{\overset{\cdot\cdot}{O}}:H\right]^+$ with H below O	$[H-O-H]^+$ with H bonded below O
(b)	OH^-	$6 + 1 + 1 = 8\ e^-$	$:\underset{\cdot\cdot}{\overset{\cdot\cdot}{O}}:H^-$	$[O-H]^-$
(c)	HS^-	$1 + 6 + 1 = 8\ e^-$	$H:\underset{\cdot\cdot}{\overset{\cdot\cdot}{S}}:^-$	$[H-S]^-$
(d)	CN^-	$4 + 5 + 1 = 10\ e^-$	$:C:::N:^-$	$[C\equiv N]^-$

Section 12.6 *Polar Covalent Bonds*

45. Within a period of elements, the electronegativity *increases from left to right* in the periodic table.

47. Nonmetals are more electronegative than metals.

49.

	More Electronegative		More Electronegative
(a)	**Cl** > Br	(b)	**O** > S
(c)	**Se** > As	(d)	N < **F**

(Note: The more electronegative element is in bold.)

51.

	Bond	Polarity		Bond	Polarity
(a)	Br—Cl	3.0 - 2.8 = 0.2	(b)	Br—F	4.0 - 2.8 = 1.2
(c)	I—Cl	3.0 - 2.5 = 0.5	(d)	I—Br	2.8 - 2.5 = 0.3

53. Polar Bonds Using Delta Notation

(a)	δ^- C—H δ^+	(b)	δ^+ Se—O δ^-
(c)	δ^+ P—I δ^-	(d)	δ^+ H—Br δ^-

(Note: δ^- indicates the more electronegative atom and δ^+ indicates the more electropositive atom.)

Section 12.7 *Nonpolar Covalent Bonds*

55.

	Bond	Polarity	Classification
(a)	Cl–Cl	3.0 – 3.0 = 0	nonpolar
(b)	Cl–N	3.0 – 3.0 = 0	nonpolar
(c)	N–H	3.0 – 2.1 = 0.9	polar
(d)	H–P	2.1 – 2.1 = 0	nonpolar

Thus, (a), (b), and (d) are nonpolar.

57. H_2, N_2, Cl_2, and Br_2 occur naturally as diatomic molecules.

Section 12.8 *Coordinate Covalent Bonds*

(Note: Coordinate covalent bonds are indicated by a dash, —.)

59.

Molecule	Valence Electrons	Electron Dot	Coord. Cov. Bond
HBrO	1 + 7 + 6 = 14 e⁻	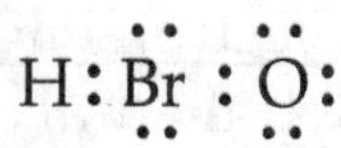	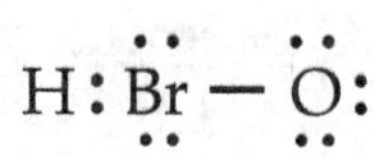

61.

Molecule	Valence Electrons	Electron Dot	Coord. Cov. Bond
$HBrO_2$	1 + 7 + 2(6) = 20 e⁻	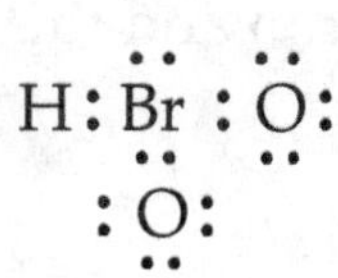	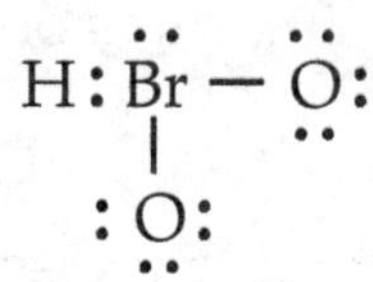

63.

Polyatomic Ion	Valence Electrons	Electron Dot	Coord. Cov. Bond
NH_4^+	5 + 4(1) – 1 = 8 e⁻	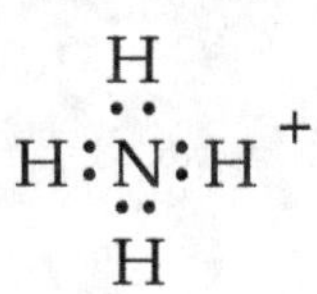	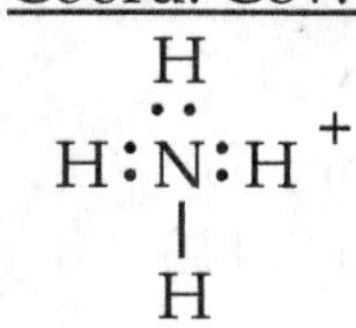

65.

Polyatomic Ion	Valence Electrons	Electron Dot	Coord. Cov. Bond
NO_3^-	5 + 3(6) + 1 = 24 e⁻	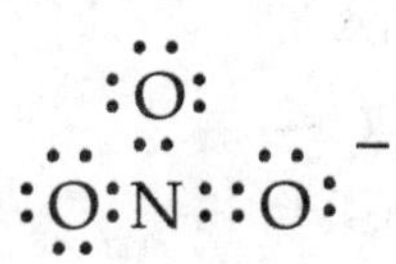	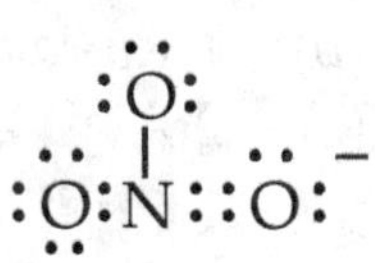

Section 12.9 *Hydrogen Bonds*

67. H–F: – – – H–F:

69. The *bond energy is much less* in a hydrogen bond than in a polar covalent bond.

Section 12.10 *Shapes of Molecules*

71.

Formula	Electron Pair	Molecular Shape	Bond Angle
SiH_4	tetrahedral	tetrahedral	109.5°

73.

Formula	Electron Pair	Molecular Shape	Bond Angle
NI_3	tetrahedral	trigonal pyramidal	107°

75.

Formula	Electron Pair	Molecular Shape	Bond Angle
H_2S	tetrahedral	bent	104.5°

77. In CF_4, each of the four **C**–F bonds is polar, but each fluorine atom pulls equally in opposite directions to give a *nonpolar molecule*. All four F atoms are at the corners of a tetrahedron and the molecular shape is described as *tetrahedral*.

General Exercises

79.

	Substance	Particle
(a)	U	atom
(b)	F_2	molecule
(c)	UF_6	formula unit
(d)	HF	molecule

81.

	Ions	Chemical Formula
(a)	3 Sr^{2+} and 2 As^{3-}	Sr_3As_2
(b)	1 Ra^{2+} and 1 O^{2-}	RaO
(c)	2 Al^{3+} and 3 $CO_3{}^{2-}$	$Al_2(CO_3)_3$
(d)	1 Cd^{2+} and 2 OH^-	$Cd(OH)_2$

3. The radius of a sodium ion is about half that of a sodium atom because the sodium ion has one less energy level (3*s*). In addition, the sodium ion has one more proton than electron, which draws the electrons closer to the nucleus.

85.

Bond	Polarity
B–Cl	3.0 – 2.0 = 1.0 (polar)

87.

Bond	Polarity
H–P	2.1 – 2.1 = 0 (nonpolar)

89.

Polar Bond and Delta Notation	
	δ^+ Ge–Cl δ^-

91.

Molecule	Valence Electrons	Electron Dot	Structural Formula
SiH_4	4 + 4(1) = 8 e⁻	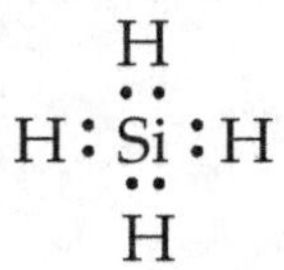	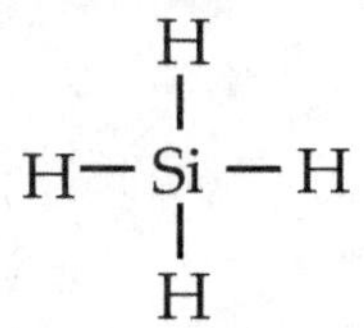

93.

Polyatomic Ion	Valence Electrons	Electron Dot	Structural Formula
AsO_3^{3-}	5 + 3(6) + 3 = 26 e⁻	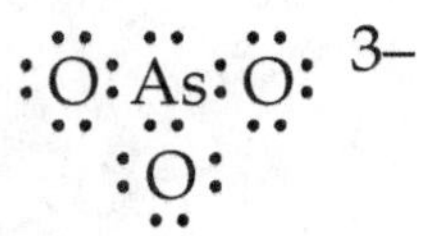	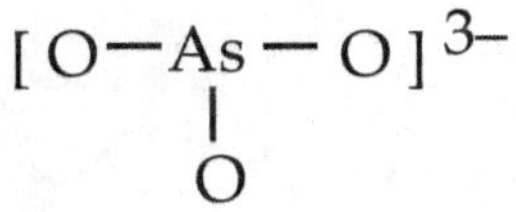

95.

Formula	Electron Pair	Molecular Shape	Bond Angle
CS_2	linear	linear	180°

97.

Formula	Electron Pair	Molecular Shape	Bond Angle
CCl_2O	trigonal planar	trigonal planar	120°

Challenge Exercises

99.

Molecule	Valence Electrons	Electron Dot	Structural Formula
XeO_2	$8 + 2(6) = 20\ e^-$	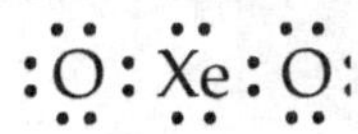	O—Xe—O

CHAPTER

Solutions 13

Section 13.1 *Gases in Solution*

1\.

	Change	Result
(a)	temperature of soda *decreases*	solubility of CO_2 gas *increases*
(b)	partial pressure of CO_2 *decreases*	solubility of CO_2 gas *decreases*

3\. solubility at 1.00 atm × pressure factor = new solubility

$$\frac{1.90 \text{ mL } N_2}{1 \text{ dL blood}} \times \frac{7.00 \cancel{\text{atm}}}{1.00 \cancel{\text{atm}}} = \frac{13.3 \text{ mL } N_2}{1 \text{ dL blood}}$$

5\.

$$1.00 \text{ atm} \times \frac{0.25 \cancel{\text{g } N_2O}}{0.12 \cancel{\text{g } N_2O}} = 2.1 \text{ atm}$$

Section 13.2 *Liquids in Solution*

7\. (a) polar solute + polar solvent = miscible
(b) polar solute + nonpolar solvent = immiscible

9\.

	Solvent	Classification
(a)	H_2O	polar
(b)	C_6H_{14}	nonpolar
(c)	C_3H_6O	polar
(d)	$CHCl_3$	nonpolar

11\.

	Solvent	Miscible or Immiscible
(a)	C_7H_{16}	miscible with C_6H_{14}
(b)	CH_3OH	immiscible with C_6H_{14}
(c)	C_4H_8O	immiscible with C_6H_{14}
(d)	$CHCl_3$	miscible with C_6H_{14}

13\. Add a few drops of water to the unknown liquid.
If the unknown liquid is miscible with water, it is *polar*.
If the unknown liquid is immiscible with water, it is *nonpolar*.

Section 13.3 *Solids in Solution*

15. (a) polar solute + nonpolar solvent = insoluble
(b) nonpolar solute + nonpolar solvent = soluble
(c) ionic solute + nonpolar solvent = insoluble

17.

	Compound	Soluble or Insoluble
(a)	naphthalene, $C_{10}H_8$	insoluble in H_2O
(b)	potassium hydroxide, KOH	soluble in H_2O
(c)	calcium acetate, $Ca(C_2H_3O_2)_2$	soluble in H_2O
(d)	trichlorotoluene, $C_7H_5Cl_3$	insoluble in H_2O

19.

	Compound	Soluble or Insoluble
(a)	oleic acid, $C_{18}H_{34}O_2$	fat soluble
(b)	citric acid, $C_6H_8O_7$	water soluble
(c)	lactic acid, $CH_3CH(OH)COOH$	water soluble
(d)	stearic acid, $CH_3(CH_2)_{16}COOH$	fat soluble

Section 13.4 *The Dissolving Process*

21. Fructose molecule, $C_6H_{12}O_6$, dissolved in water:

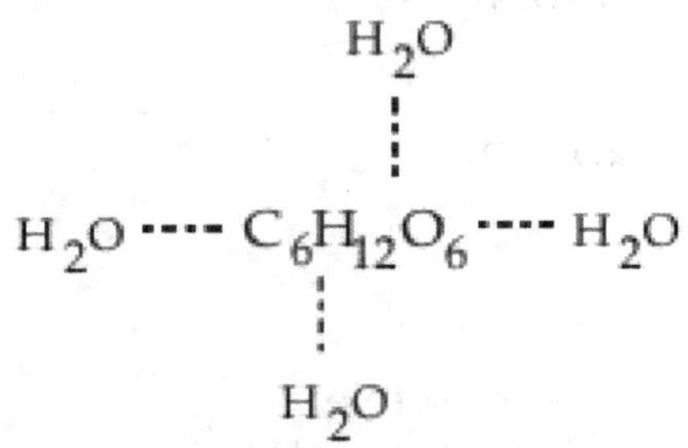

23. (a) Lithium bromide, LiBr, dissolved in water:

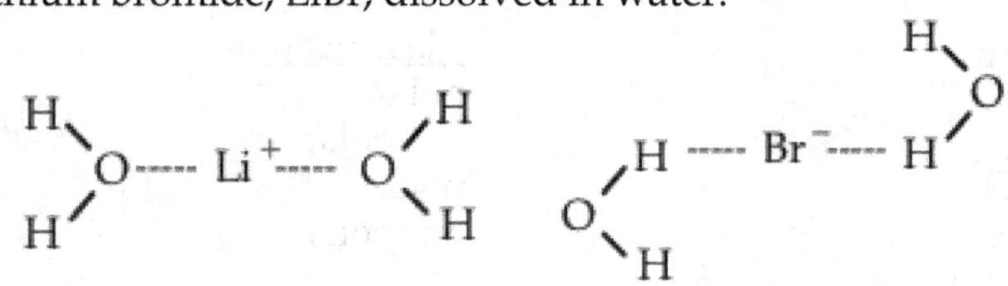

(b) Calcium chloride, $CaCl_2$, dissolved in water:

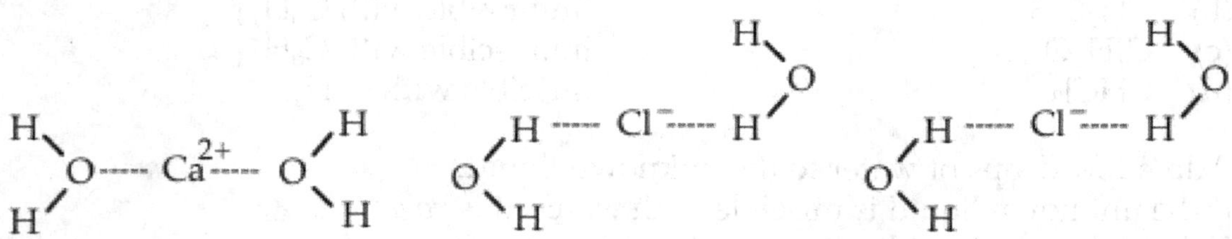

Section 13.5 *Rate of Dissolving*

25. The three factors that increase the rate of dissolving a solid solute in solution:
    (1) heating the solution
    (2) stirring the solution
    (3) grinding the solute

Section 13.6 *Solubility and Temperature*

27. (a) ~35 g NaCl/100 g H_2O (b) ~35 g KCl/100 g H_2O

29. (a) ~35 g NaCl/100 g H_2O (b) ~40 g KCl/100 g H_2O

31. (a) ~0 °C (b) ~70 °C

33. (a) ~100 °C (b) ~35 °C

35. (a) supersaturated at 50 °C
    (b) saturated at 70 °C
    (c) unsaturated at 90 °C

Section 13.7 *Unsaturated, Saturated, and Supersaturated Solutions*

37. The solubility of LiCl is ~80 g/100 g water at 20 °C (see textbook Figure 13.5).
    (a) ~80 g LiCl remains in solution
    (b) ~20 g LiCl (100 g – 80 g) crystallizes from solution

39. (a) supersaturated at 70 °C
    (b) saturated at 80 °C
    (c) unsaturated at 90 °C

41. $25 \cancel{\text{g H}_2\text{O}} \times \frac{40 \text{ g rock salt}}{100 \cancel{\text{g H}_2\text{O}}} = 10 \text{ g rock salt can dissolve}$
    The solution contains 10 g of rock salt; thus, the solution is *saturated* at 30 °C.

Section 13.8 *Mass/Mass Percent Concentration*

43. $$\frac{\text{mass of solute}}{\text{mass of solution}} \times 100\% = \text{m/m }\%$$

(a) $$\frac{1.25 \text{ g NaCl}}{100.0 \text{ g solution}} \times 100\% = 1.25\%$$

(b) $$\frac{2.50 \text{ g } K_2Cr_2O_7}{95.0 \text{ g solution}} \times 100\% = 2.63\%$$

(c) $$\frac{10.0 \text{ g } CaCl_2}{250.0 \text{ g solution}} \times 100\% = 4.00\%$$

(d) $$\frac{65.0 \text{ g sugar}}{125.0 \text{ g solution}} \times 100\% = 52.0\%$$

45.

	Solution	Unit Factors		
(a)	1.50% KBr	$\frac{1.50 \text{ g KBr}}{100 \text{ g solution}}$	and	$\frac{100 \text{ g solution}}{1.50 \text{ g KBr}}$
		$\frac{98.50 \text{ g } H_2O}{100 \text{ g solution}}$	and	$\frac{100 \text{ g solution}}{98.50 \text{ g } H_2O}$
		$\frac{1.50 \text{ g KBr}}{98.50 \text{ g } H_2O}$	and	$\frac{98.50 \text{ g } H_2O}{1.50 \text{ g KBr}}$
(b)	2.50% $AlCl_3$	$\frac{2.50 \text{ g } AlCl_3}{100 \text{ g solution}}$	and	$\frac{100 \text{ g solution}}{2.50 \text{ g } AlCl_3}$
		$\frac{97.50 \text{ g } H_2O}{100 \text{ g solution}}$	and	$\frac{100 \text{ g solution}}{97.50 \text{ g } H_2O}$
		$\frac{2.50 \text{ g } AlCl_3}{97.50 \text{ g } H_2O}$	and	$\frac{97.50 \text{ g } H_2O}{2.50 \text{ g } AlCl_3}$
(c)	3.75% $AgNO_3$	$\frac{3.75 \text{ g } AgNO_3}{100 \text{ g solution}}$	and	$\frac{100 \text{ g solution}}{3.75 \text{ g } AgNO_3}$
		$\frac{96.25 \text{ g } H_2O}{100 \text{ g solution}}$	and	$\frac{100 \text{ g solution}}{96.25 \text{ g } H_2O}$
		$\frac{3.75 \text{ g } AgNO_3}{96.25 \text{ g } H_2O}$	and	$\frac{96.25 \text{ g } H_2O}{3.75 \text{ g } AgNO_3}$

(d) 4.25% Li_2SO_4 $\quad \dfrac{4.25 \text{ g } Li_2SO_4}{100 \text{ g solution}}$ and $\dfrac{100 \text{ g solution}}{4.25 \text{ g } Li_2SO_4}$

$\dfrac{95.75 \text{ g } H_2O}{100 \text{ g solution}}$ and $\dfrac{100 \text{ g solution}}{95.75 \text{ g } H_2O}$

$\dfrac{4.25 \text{ g } Li_2SO_4}{95.75 \text{ g } H_2O}$ and $\dfrac{95.75 \text{ g } H_2O}{4.25 \text{ g } Li_2SO_4}$

47. (a) $5.36 \cancel{\text{g glucose}} \times \dfrac{100 \text{ g solution}}{10.0 \cancel{\text{g glucose}}} = 53.6 \text{ g solution}$

(b) $25.0 \cancel{\text{g sucrose}} \times \dfrac{100 \text{ g solution}}{12.5 \cancel{\text{g sucrose}}} = 200 \text{ g solution } (2.00 \times 10^2 \text{ g})$

49. (a) $85.0 \cancel{\text{g solution}} \times \dfrac{2.00 \text{ g } FeBr_2}{100 \cancel{\text{g solution}}} = 1.70 \text{ g } FeBr_2$

(b) $105.0 \cancel{\text{g solution}} \times \dfrac{5.00 \text{ g } Na_2CO_3}{100 \cancel{\text{g solution}}} = 5.25 \text{ g } Na_2CO_3$

51. (a) $100.0 \cancel{\text{g solution}} \times \dfrac{95.0 \text{ g } H_2O}{100 \cancel{\text{g solution}}} = 95.0 \text{ g } H_2O$

(b) $250.0 \cancel{\text{g solution}} \times \dfrac{90.0 \text{ g } H_2O}{100 \cancel{\text{g solution}}} = 225 \text{ g } H_2O$

Section 13.9 *Molar Concentration*

53. (a) MM of NaCl $= 22.99 \text{ g/mol} + 35.45 \text{ g/mol}$
$= 58.44 \text{ g/mol}$

$$\frac{1.50 \cancel{\text{g NaCl}}}{100.0 \cancel{\text{mL solution}}} \times \frac{1 \text{ mol NaCl}}{58.45 \cancel{\text{g NaCl}}} \times \frac{1000 \cancel{\text{mL solution}}}{1 \text{ L solution}} = 0.257 \, M \text{ NaCl}$$

(b) MM of $K_2Cr_2O_7$ $= 2(39.10 \text{ g/mol}) + 2(52.00 \text{ g/mol}) + 7(16.00 \text{ g/mol})$
$= 294.20 \text{ g/mol}$

$$\frac{1.50 \cancel{\text{g } K_2Cr_2O_7}}{100.0 \cancel{\text{mL solution}}} \times \frac{1 \text{ mol } K_2Cr_2O_7}{294.20 \cancel{\text{g } K_2Cr_2O_7}} \times \frac{1000 \cancel{\text{mL solution}}}{1 \text{ L solution}} = 0.0510 \, M \; K_2Cr_2O_7$$

(c) MM of $CaCl_2$ = 40.08 g/mol + 2(35.45 g/mol)
= 110.98 g/mol

$$\frac{5.55 \text{ g } CaCl_2}{125 \text{ mL solution}} \times \frac{1 \text{ mol } CaCl_2}{110.98 \text{ g } CaCl_2} \times \frac{1000 \text{ mL solution}}{1 \text{ L solution}} = 0.400\ M\ CaCl_2$$

(d) MM of Na_2SO_4 = 2(22.99 g/mol) + 32.07 g/mol + 4(16.00 g/mol)
= 142.05 g/mol

$$\frac{5.55 \text{ g } Na_2SO_4}{125 \text{ mL solution}} \times \frac{1 \text{ mol } Na_2SO_4}{142.05 \text{ g } Na_2SO_4} \times \frac{1000 \text{ mL solution}}{1 \text{ L solution}} = 0.313\ M\ Na_2SO_4$$

55.

	Solution	Unit Factors		
(a)	0.100 *M* LiI	$\frac{0.100 \text{ mol LiI}}{1 \text{ L solution}}$	and	$\frac{1 \text{ L solution}}{0.100 \text{ mol LiI}}$
		$\frac{0.100 \text{ mol LiI}}{1000 \text{ mL solution}}$	and	$\frac{1000 \text{ mL solution}}{0.100 \text{ mol LiI}}$
(b)	0.100 *M* $NaNO_3$	$\frac{0.100 \text{ mol } NaNO_3}{1 \text{ L solution}}$	and	$\frac{1 \text{ L solution}}{0.100 \text{ mol } NaNO_3}$
		$\frac{0.100 \text{ mol } NaNO_3}{1000 \text{ mL solution}}$	and	$\frac{1000 \text{ mL solution}}{0.100 \text{ mol } NaNO_3}$
(c)	0.500 *M* K_2CrO_4	$\frac{0.500 \text{ mol } K_2CrO_4}{1 \text{ L solution}}$	and	$\frac{1 \text{ L solution}}{0.500 \text{ mol } K_2CrO_4}$
		$\frac{0.500 \text{ mol } K_2CrO_4}{1000 \text{ mL solution}}$	and	$\frac{1000 \text{ mL solution}}{0.500 \text{ mol } K_2CrO_4}$
(d)	0.500 *M* $ZnSO_4$	$\frac{0.500 \text{ mol } ZnSO_4}{1 \text{ L solution}}$	and	$\frac{1 \text{ L solution}}{0.500 \text{ mol } ZnSO_4}$
		$\frac{0.500 \text{ mol } ZnSO_4}{1000 \text{ mL solution}}$	and	$\frac{1000 \text{ mL solution}}{0.500 \text{ mol } ZnSO_4}$

57. (a) MM of KNO_3 = 39.10 g/mol + 14.01 g/mol + 3(16.00 g/mol) = 101.11 g/mol

$$2.50\ \cancel{g\ KNO_3} \times \frac{1\ \cancel{mol\ KNO_3}}{101.11\ \cancel{g\ KNO_3}} \times \frac{1\ L\ solution}{0.325\ \cancel{mol\ KNO_3}} = 0.0761\ L\ solution$$

(b) MM of $AlBr_3$ = 26.98 g/mol + 3(79.90 g/mol) = 266.68 g/mol

$$2.50\ \cancel{g\ AlBr_3} \times \frac{1\ \cancel{mol\ AlBr_3}}{266.68\ \cancel{g\ AlBr_3}} \times \frac{1\ L\ solution}{0.325\ \cancel{mol\ AlBr_3}} = 0.0288\ L\ solution$$

(c) MM of $Co(C_2H_3O_2)_2$ = 58.93 g/mol + 4(12.01 g/mol) + 6(1.01 g/mol) + 4(16.00 g/mol) = 177.03 g/mol

$$2.50\ \cancel{g\ Co(C_2H_3O_2)_2} \times \frac{1\ \cancel{mol\ Co(C_2H_3O_2)_2}}{177.03\ \cancel{g\ Co(C_2H_3O_2)_2}} \times \frac{1\ L\ solution}{1.00\ \cancel{mol\ Co(C_2H_3O_2)_2}} = 0.0141\ L\ solution$$

(d) MM of $(NH_4)_3PO_4$ = 3(14.01 g/mol) + 12(1.01 g/mol) + 30.97 g/mol + 4(16.00 g/mol) = 149.12 g/mol

$$2.50\ \cancel{g\ (NH_4)_3PO_4} \times \frac{1\ \cancel{mol\ (NH_4)_3PO_4}}{149.12\ \cancel{g\ (NH_4)_3PO_4}} \times \frac{1\ L\ solution}{1.00\ \cancel{mol\ (NH_4)_3PO_4}} = 0.0168\ L\ solution$$

59. (a) MM of $FeCl_3$ = 55.85 g/mol + 3(35.45 g/mol) = 162.20 g/mol

$$2.25\ \cancel{L\ solution} \times \frac{0.200\ \cancel{mol\ FeCl_3}}{1\ \cancel{L\ solution}} \times \frac{162.20\ g\ FeCl_3}{1\ \cancel{mol\ FeCl_3}} = 73.0\ g\ FeCl_3$$

(b) MM of KIO_4 = 39.10 g/mol + 126.90 g/mol + 4(16.00 g/mol) = 230.00 g/mol

$$2.25\ \cancel{L\ solution} \times \frac{0.200\ \cancel{mol\ KIO_4}}{1\ \cancel{L\ solution}} \times \frac{230.00\ g\ KIO_4}{1\ \cancel{mol\ KIO_4}} = 104\ g\ KIO_4$$

(c) MM of $ZnSO_4$ = 65.39 g/mol + 32.07 g/mol + 4(16.00 g/mol) = 161.46 g/mol

$$50.0\ \cancel{mL\ solution} \times \frac{0.295\ \cancel{mol\ ZnSO_4}}{1000\ \cancel{mL\ solution}} \times \frac{161.46\ g\ ZnSO_4}{1\ \cancel{mol\ ZnSO_4}} = 2.38\ g\ ZnSO_4$$

(d) MM of $Ni(NO_3)_2$ = 58.69 g/mol + 2(14.01 g/mol) + 6(16.00 g/mol)
= 182.71 g/mol

$$50.0\ \cancel{\text{mL solution}} \times \frac{0.295\ \cancel{\text{mol Ni(NO}_3)_2}}{1000\ \cancel{\text{mL solution}}} \times \frac{182.71\ \text{g Ni(NO}_3)_2}{1\ \cancel{\text{mol Ni(NO}_3)_2}} = 2.69\ \text{g Ni(NO}_3)_2$$

61. MM of $Ca(OH)_2$ = 40.08 g/mol + 2(16.00 g/mol) + 2(1.01 g/mol)
= 74.10 g/mol

$$\frac{0.185\ \cancel{\text{g Ca(OH)}_2}}{100.0\ \cancel{\text{mL solution}}} \times \frac{1\ \text{mol Ca(OH)}_2}{74.10\ \cancel{\text{g Ca(OH)}_2}} \times \frac{1000\ \cancel{\text{mL solution}}}{1\ \text{L solution}} = 0.0250\ M\ \text{Ca(OH)}_2$$

63. (a) $$\frac{0.524\ \text{g glucose}}{10.483\ \text{g solution}} \times 100\% = 5.00\%$$

(b) MM of $C_6H_{12}O_6$ = 6(12.01 g/mol) + 12(1.01 g/mol) + 6(16.00 g/mol)
= 180.18 g/mol

$$\frac{0.524\ \cancel{\text{g C}_6\text{H}_{12}\text{O}_6}}{100.0\ \cancel{\text{mL solution}}} \times \frac{1\ \text{mol C}_6\text{H}_{12}\text{O}_6}{180.18\ \cancel{\text{g C}_6\text{H}_{12}\text{O}_6}} \times \frac{1000\ \cancel{\text{mL solution}}}{1\ \text{L solution}} = 0.291\ M\ \text{C}_6\text{H}_{12}\text{O}_6$$

Section 13.10 *Dilution of a Solution*

65. $$10.0\ \cancel{\text{mL solution}} \times \frac{12\ \text{moL HCl}}{1000\ \cancel{\text{mL solution}}} = 0.12\ \text{mol HCl}$$

$$\frac{0.12\ \text{mol HCl}}{250.0\ \cancel{\text{mL solution}}} \times \frac{1000\ \cancel{\text{mL solution}}}{1\ \text{L solution}} = \frac{0.48\ \text{mol HCl}}{1\ \text{L solution}} = 0.48\ M\ \text{HCl}$$

Algebraic Solution:

$$M_1 \times V_1 = M_2 \times V_2$$

$$\frac{12\ M \times 10.0\ \cancel{\text{mL}}}{250.0\ \cancel{\text{mL}}} = 0.48\ M\ \text{HCl}$$

67. $2.50 \text{ L solution} \times \dfrac{0.10 \text{ moL } H_2SO_4}{1 \text{ L solution}} = 0.25 \text{ mol } H_2SO_4$

$0.25 \text{ mol } H_2SO_4 \times \dfrac{1 \text{ L solution}}{18 \text{ mol } H_2SO_4} = 0.014 \text{ L } H_2SO_4$

Algebraic Solution:

$$M_1 \times V_1 = M_2 \times V_2$$

$$\frac{0.10\,M \times 2.50 \text{ L}}{18\,M} = 0.014 \text{ L } H_2SO_4 \text{ (14 mL } H_2SO_4)$$

Section 13.11 *Solution Stoichiometry*

69. $MgI_2(aq) + 2\,AgNO_3(aq) \rightarrow 2\,AgI(s) + Mg(NO_3)_2(aq)$

(a) $50.0 \text{ mL } AgNO_3 \times \dfrac{0.125 \text{ mol } AgNO_3}{1000 \text{ mL } AgNO_3} \times \dfrac{1 \text{ mol } Mg(NO_3)_2}{2 \text{ mol } AgNO_3}$

$= 0.00313 \text{ mol } Mg(NO_3)_2$

$\dfrac{0.00313 \text{ mol } Mg(NO_3)_2}{15.0 \text{ mL solution}} \times \dfrac{1000 \text{ mL solution}}{1 \text{ L solution}} = 0.208\,M \; Mg(NO_3)_2$

(b) $50.0 \text{ mL } AgNO_3 \times \dfrac{0.125 \text{ mol } AgNO_3}{1000 \text{ mL } AgNO_3} \times \dfrac{1 \text{ mol AgI}}{1 \text{ mol } AgNO_3} \times \dfrac{234.77 \text{ g AgI}}{1 \text{ mol AgI}}$

$= 1.47 \text{ g AgI}$

General Exercises

71. Fog demonstrates the Tyndall effect when it contains small water droplets ranging in size from 1–100 nm.

73. A helium gas mixture is used rather than compressed air for two reasons:
(1) Helium is less soluble than nitrogen in blood.
(2) Helium atoms escape from blood more rapidly than nitrogen molecules; thus, allowing for quicker decompression.

75. 100 mL solution = 0.100 L solution

$1.00 \text{ L solution} \times \dfrac{22.8 \text{ g } SO_2}{0.100 \text{ L solution}} = 228 \text{ g } SO_2$

77. Although ethyl ether, $C_4H_{10}O$, contains two C–O polar bonds, it is only partially miscible in water because it has mostly nonpolar C–H bonds.

79. Water can hydrogen bond with the polar –OH in an alcohol, and draw it into solution. As the nonpolar C_xH_y– portion of the alcohol molecule increases, the alcohol becomes less soluble, and eventually immiscible with water.

81. (a) In a 40% solution, ethanol is the solute and water is the solvent.
(b) In a 95% solution, water is the solute and ethanol is the solvent.

Challenge Exercises

83. MM of CH_3COOH = 2(12.01 g/mol) + 4(1.01 g/mol) + 2(16.00 g/mol)

= 60.06 g/mol

5.00% CH_3COOH = 5.00 g CH_3COOH per 100 g vinegar

$$\text{Solute: } 5.00\ \cancel{\text{g } CH_3COOH} \times \frac{1 \text{ mol } CH_3COOH}{60.06\ \cancel{\text{g } CH_3COOH}} = 0.0833 \text{ mol } CH_3COOH$$

$$\text{Solution: } 100\ \cancel{\text{g vinegar}} \times \frac{1\ \cancel{\text{mL vinegar}}}{1.01\ \cancel{\text{g vinegar}}} \times \frac{1 \text{ L vinegar}}{1000\ \cancel{\text{mL vinegar}}}$$

$$= 0.0990 \text{ L vinegar}$$

$$\text{Molarity: } \frac{0.0833 \text{ mol } HC_2H_3O_2}{0.0990 \text{ L vinegar}} = 0.841\ M\ CH_3COOH$$

CHAPTER

14

Acids and Bases

Section 14.1 *Properties of Acids and Bases*

1. General Properties of an Acidic Solution
   (a) *True*, an acidic solution has a sour taste.
   (b) *True*, an acidic solution turns litmus paper red.
   (c) *False*, an acidic solution has a pH less than 7.
   (d) *False*, a basic solution feels slippery.

3.

	Aqueous Solution	pH	Classification
(a)	stomach acid	1.5	strongly acidic
(b)	vinegar	3.2	weakly acidic
(c)	distilled water	7.0	neutral
(d)	drain cleaner	13.5	strongly basic

Section 14.2 *Arrhenius Acids and Bases*

5.

	Acid	Ionization	Strength
(a)	$HClO(aq)$	~1%	weak acid
(b)	$HClO_2(aq)$	~1%	weak acid
(c)	$HClO_3(aq)$	~100%	strong acid
(d)	$HClO_4(aq)$	~100%	strong acid

7.

	Formula	Classification		Formula	Classification
(a)	$H_2SO_4(aq)$	Arrhenius acid	(b)	$Na_2SO_4(aq)$	salt
(c)	$NaOH(aq)$	Arrhenius base	(d)	$Na_2CO_3(aq)$	salt

9.

	Arrhenius Acid	Arrhenius Base
(a)	$HBr(aq)$	$NaOH(aq)$
(b)	$H_2SO_4(aq)$	$Mg(OH)_2(aq)$

11.

	Salt	Acid	Base
(a)	$NaF(aq)$	HF	$NaOH$
(b)	$MgI_2(aq)$	HI	$Mg(OH)_2$
(c)	$Ca(NO_3)_2(aq)$	HNO_3	$Ca(OH)_2$
(d)	$Li_2CO_3(aq)$	H_2CO_3	$LiOH$

Neutralization Reactions:

(a) $HF(aq) + NaOH(aq) \rightarrow NaF(aq) + H_2O(l)$
(b) $2\,HI(aq) + Mg(OH)_2(s) \rightarrow MgI_2(aq) + 2\,H_2O(l)$
(c) $2\,HNO_3(aq) + Ca(OH)_2(aq) \rightarrow Ca(NO_3)_2(aq) + 2\,H_2O(l)$
(d) $H_2CO_3(aq) + 2\,LiOH(aq) \rightarrow Li_2CO_3(aq) + 2\,H_2O(l)$

13. (a) $2\,HNO_3(aq) + Ca(OH)_2(aq) \rightarrow Ca(NO_3)_2(aq) + 2\,H_2O(l)$
(b) $H_2CO_3(aq) + Ba(OH)_2(aq) \rightarrow BaCO_3(aq) + 2\,H_2O(l)$

Section 14.3 *Brønsted–Lowry Acids and Bases*

15.

	Acid	Base		Acid	Base
(a)	$HC_2H_3O_2(aq)$	$LiOH(aq)$	(b)	$HCl(aq)$	$NH_3(aq)$

17.

	Acid	Base		Acid	Base
(a)	$HI(aq)$	$H_2O(l)$	(b)	$HC_2H_3O_2(aq)$	$HS^-(aq)$

19. (a) $HF(aq) + NaHS(aq) \rightarrow H_2S(aq) + NaF(aq)$
(b) $HNO_2(aq) + NaC_2H_3O_2(aq) \rightarrow NaNO_2(aq) + HC_2H_3O_2(aq)$

Section 14.4 *Acid–Base Indicators*

21.

	pH	Phenolphthalein Color		pH	Phenolphthalein Color
(a)	6	colorless	(b)	12	pink

23.

	pH	Methyl Red Color		pH	Methyl Red Color
(a)	3	red	(b)	7	yellow

25.

	pH	Bromthymol Blue Color		pH	Bromthymol Blue Color
(a)	5	yellow	(b)	9	blue

Section 14.5 *Acid–Base Titrations*

27. $2\,HNO_3(aq) + Ba(OH)_2(aq) \rightarrow Ba(NO_3)_2(aq) + 2\,H_2O(l)$

$$21.50\ \cancel{\text{mL solution}} \times \frac{0.115\ \text{mol Ba(OH)}_2}{1000\ \cancel{\text{mL solution}}} = 0.00247\ \text{mol Ba(OH)}_2$$

$$0.00247\ \cancel{\text{mol Ba(OH)}_2} \times \frac{2\ \text{mol HNO}_3}{1\ \cancel{\text{mol Ba(OH)}_2}} = 0.00495\ \text{mol HNO}_3$$

$$\frac{0.00495\ \text{mol HNO}_3}{25.0\ \cancel{\text{mL solution}}} \times \frac{1000\ \cancel{\text{mL solution}}}{1\ \text{L solution}} = \frac{0.198\ \text{mol HNO}_3}{1\ \text{L solution}}$$

Molarity of nitric acid: 0.198 M HNO_3

29. $H_2SO_4(aq) + 2\,NH_4OH(aq) \rightarrow Na_2SO_4(aq) + 2\,H_2O(l)$

$$28.15\ \cancel{\text{mL solution}} \times \frac{0.100\ \text{mol NH}_4\text{OH}}{1000\ \cancel{\text{mL solution}}} = 0.00282\ \text{mol NH}_4\text{OH}$$

$$0.00282\ \cancel{\text{mol}}\ \text{NH}_4\cancel{\text{OH}} \times \frac{1\ \text{mol H}_2\text{SO}_4}{2\ \cancel{\text{mol}}\ \text{NH}_4\cancel{\text{OH}}} = 0.00141\ \text{mol H}_2\text{SO}_4$$

$$\frac{0.00141\ \text{mol H}_2\text{SO}_4}{10.0\ \cancel{\text{mL solution}}} \times \frac{1000\ \cancel{\text{mL solution}}}{1\ \text{L solution}} = \frac{0.141\ \text{mol H}_2\text{SO}_4}{1\ \text{L solution}}$$

Molarity of sulfuric acid: 0.141 M H_2SO_4

31. $H_3PO_4(aq) + 3\,LiOH(aq) \rightarrow Li_3PO_4(aq) + 3\,H_2O(l)$

$$42.45\ \cancel{\text{mL solution}} \times \frac{0.500\ \text{mol LiOH}}{1000\ \cancel{\text{mL solution}}} = 0.0212\ \text{mol}\ \cancel{\text{Li}}\text{OH}$$

$$0.0212\ \cancel{\text{mol LiOH}} \times \frac{1\ \text{mol H}_3\text{PO}_4}{3\ \cancel{\text{mol LiOH}}} = 0.00708\ \text{mol H}_3\text{PO}_4$$

$$\frac{0.00708\ \text{mol H}_3\text{PO}_4}{50.0\ \cancel{\text{mL solution}}} \times \frac{1000\ \cancel{\text{mL solution}}}{1\ \text{L solution}} = \frac{0.142\ \text{mol H}_3\text{PO}_4}{1\ \text{L solution}}$$

Molarity of phosphoric acid: 0.142 M H_3PO_4

33. (a) MM of HCl = 36.46 g/mol

$$\frac{6.00 \cancel{\text{mol HCl}}}{1000 \cancel{\text{mL solution}}} \times \frac{36.46 \text{ g HCl}}{1 \cancel{\text{mol HCl}}} \times \frac{1 \cancel{\text{mL solution}}}{1.10 \text{ g solution}} \times 100\%$$

$$= 19.9\% \text{ HCl}$$

(b) MM of $HC_2H_3O_2$ = 60.06 g/mol

$$\frac{1.00 \cancel{\text{mol HC}_2\text{H}_3\text{O}_2}}{1000 \cancel{\text{mL solution}}} \times \frac{60.06 \text{ g HC}_2\text{H}_3\text{O}_2}{1 \cancel{\text{mol HC}_2\text{H}_3\text{O}_2}} \times \frac{1 \cancel{\text{mL solution}}}{1.01 \text{ g solution}} \times 100\%$$

$$= 5.95\% \text{ HC}_2\text{H}_3\text{O}_2$$

(c) MM of HNO_3 = 63.02 g/mol

$$\frac{0.500 \cancel{\text{mol HNO}_3}}{1000 \cancel{\text{mL solution}}} \times \frac{63.02 \text{ g HNO}_3}{1 \cancel{\text{mol HNO}_3}} \times \frac{1 \cancel{\text{mL solution}}}{1.01 \text{ g solution}} \times 100\%$$

$$= 3.12\% \text{ HNO}_3$$

(d) MM of H_2SO_4 = 98.09 g/mol

$$\frac{3.00 \cancel{\text{mol H}_2\text{SO}_4}}{1000 \cancel{\text{mL solution}}} \times \frac{98.09 \text{ g H}_2\text{SO}_4}{1 \cancel{\text{mol H}_2\text{SO}_4}} \times \frac{1 \cancel{\text{mL solution}}}{1.18 \text{ g solution}} \times 100\%$$

$$= 24.9\% \text{ H}_2\text{SO}_4$$

Section 14.6 *Acid–Base Standardization*

35. $HCl(aq) + NaHCO_3(aq) \rightarrow NaCl(aq) + H_2O(l) + CO_2(g)$

MM of $NaHCO_3$ = 84.01 g/mol

$$1.115 \cancel{\text{g NaHCO}_3} \times \frac{1 \cancel{\text{mol NaHCO}_3}}{84.01 \cancel{\text{g NaHCO}_3}} \times \frac{1 \text{ mol HCl}}{1 \cancel{\text{mol NaHCO}_3}} = 0.01327 \text{ mol HCl}$$

$$\frac{0.01327 \text{ mol HCl}}{33.25 \cancel{\text{mL solution}}} \times \frac{1000 \cancel{\text{mL solution}}}{1 \text{ L solution}} = \frac{0.3992 \text{ mol HCl}}{1 \text{ L solution}}$$

Molarity of hydrochloric acid: 0.3992 *M* HCl

37. $H_2C_2O_4(aq) + 2\,LiOH(aq) \rightarrow Li_2C_2O_4(aq) + 2\,H_2O(l)$

MM of $H_2C_2O_4$ = 90.04 g/mol

$$0.506\ \cancel{g\ H_2C_2O_4} \times \frac{1\ \cancel{mol\ H_2C_2O_4}}{90.04\ \cancel{g\ H_2C_2O_4}} \times \frac{2\ mol\ LiOH}{1\ \cancel{mol\ H_2C_2O_4}} = 0.0112\ mol\ LiOH$$

$$\frac{0.0112\ mol\ LiOH}{28.85\ \cancel{mL\ solution}} \times \frac{1000\ \cancel{mL\ solution}}{1\ L\ solution} = \frac{0.390\ mol\ LiOH}{1\ L\ solution}$$

Molarity of lithium hydroxide: 0.390 *M* LiOH

39. $2\,KHC_8H_4O_4(aq) + Ca(OH)_2(aq) \rightarrow CaK_2(C_8H_4O_4)_2(aq + 2\,H_2O(l)$

MM of $KHC_8H_4O_4$ = 204.23 g/mol

$$1.466\ \cancel{g\ KHC_8H_4O_4} \times \frac{1\ \cancel{mol\ KHC_8H_4O_4}}{204.23\ \cancel{g\ KHC_8H_4O_4}} \times \frac{1\ mol\ Ca(OH)_2}{2\ \cancel{mol\ KHC_8H_4O_4}}$$

$$= 0.003589\ mol\ Ca(OH)_2$$

$$0.003589\ \cancel{mol\ Ca(OH)_2} \times \frac{1000\ mL\ solution}{0.128\ \cancel{mol\ Ca(OH)_2}} = 28.0\ mL\ solution$$

41. $HAsc(aq) + NaOH(aq) \rightarrow NaAsc(aq) + H_2O(l)$

$$30.95\ \cancel{mL\ solution} \times \frac{0.176\ \cancel{mol\ NaOH}}{1000\ \cancel{mL\ solution}} \times \frac{1\ mol\ HAsc}{1\ \cancel{mol\ NaOH}} = 0.00545\ mol\ HAsc$$

$$MM\ of\ vitamin\ C = \frac{0.959\ g\ HAsc}{0.00545\ mol\ HAsc} = 176\ g/mol$$

Section 14.7 *Ionization of Water*

43. (a) Simplified ionization equation: $H_2O(l) \rightarrow H^+(aq) + OH^-(aq)$
(b) Ionization constant expression: $K_w = [H^+]\ [OH^-]$
(c) Ionization constant for water at 25 °C: $K_w = 1.0 \times 10^{-14}$

45. $K_w = [H^+][OH^-] = 1.0 \times 10^{-14}$

(a) $[OH^-] = \dfrac{1.0 \times 10^{-14}}{0.025} = 4.0 \times 10^{-13}\ M$

(b) $[OH^-] = \dfrac{1.0 \times 10^{-14}}{0.000\,017} = 5.9 \times 10^{-10}\ M$

47. $K_w = [H^+][OH^-] = 1.0 \times 10^{-14}$

(a) $[H^+] = \dfrac{1.0 \times 10^{-14}}{0.0016} = 6.3 \times 10^{-12}\ M$

(b) $[H^+] = \dfrac{1.0 \times 10^{-14}}{0.000\,29} = 3.4 \times 10^{-11}\ M$

Section 14.8 *The pH Concept*

49. $pH = -\log [H^+]$

(a) $pH = -\log 0.001 = -\log 10^{-3}$
$pH = -(-3) = 3$

(b) $pH = -\log 0.000\,01 = -\log 10^{-5}$
$pH = -(-5) = 5$

51. $[H^+] = 10^{-pH}$

(a) $[H^+] = 10^{-6} = 0.000\,001\ M\ (1 \times 10^{-6}\ M)$

(b) $[H^+] = 10^{-8} = 0.000\,000\,01\ M\ (1 \times 10^{-8}\ M)$

Section 14.9 *Advanced pH Calculations*

53. (a) $pH = -\log 0.000\,000\,30 = -\log 3.0 \times 10^{-7}$
$pH = -\log 3.0 - \log 10^{-7}$
$pH = -0.48 - (-7) = 6.52$

(b) $pH = -\log 0.000\,000\,016 = -\log 1.6 \times 10^{-8}$
$pH = -\log 1.6 - \log 10^{-8}$
$pH = -0.20 - (-8) = 7.80$

55. (a) $[H^+] = 10^{-6.55} = 10^{0.45} \times 10^{-7}$
$[H^+] = 2.8 \times 10^{-7}\ M = 0.000\,000\,28\ M$

(b) $[H^+] = 10^{-7.50} = 10^{0.50} \times 10^{-8}$
$[H^+] = 3.2 \times 10^{-8}\ M = 0.000\,000\,032\ M$

57. (a) $[OH^-] = 0.11\ M$

$$[H^+] = \frac{1.0 \times 10^{-14}}{[OH^-]} = \frac{1.0 \times 10^{-14}}{0.11} = 9.1 \times 10^{-14}$$

$pH = -\log(9.1 \times 10^{-14})$
$pH = -\log 9.1 - \log 10^{-14}$
$pH = -0.96 - (-14) = 13.04$

(b) $[OH^-] = 0.000\,55\ M = 5.5 \times 10^{-4}\ M$

$$[H^+] = \frac{1.0 \times 10^{-14}}{[OH^-]} = \frac{1.0 \times 10^{-14}}{5.5 \times 10^{-4}} = 1.8 \times 10^{-11}\ M$$

$pH = -\log(1.8 \times 10^{-11})$
$pH = -\log 1.8 - \log 10^{-11}$
$pH = -0.26 - (-11) = 10.74$

59. (a) $pH = 4.55$
$[H^+] = 10^{-4.55} = 10^{0.45} \times 10^{-5}$
$[H^+] = 2.8 \times 10^{-5}\ M$

$$[OH^-] = \frac{1.0 \times 10^{-14}}{[H^+]} = \frac{1.0 \times 10^{-14}}{2.8 \times 10^{-5}} = 3.6 \times 10^{-10}\ M$$

(b) $pH = 5.20$
$[H^+] = 10^{-5.20} = 10^{0.80} \times 10^{-6}$
$[H^+] = 6.3 \times 10^{-6}\ M$

$$[OH^-] = \frac{1.0 \times 10^{-14}}{[H^+]} = \frac{1.0 \times 10^{-14}}{6.3 \times 10^{-6}} = 1.6 \times 10^{-9}\ M$$

Section 14.10 *Strong and Weak Electrolytes*

61.

	Solution	Ionization
(a)	strong acids	highly ionized
(b)	strong bases	highly ionized
(c)	soluble ionic compounds	highly ionized

63.

	Solution	Electrolyte
(a)	$HF(aq)$	weak
(b)	$HCl(aq)$	strong
(c)	$HNO_3(aq)$	strong
(d)	$HNO_2(aq)$	weak

65.

	Solution	Electrolyte
(a)	$ZnCO_3(s)$	weak
(b)	$Sr(NO_3)_2(aq)$	strong
(c)	$K_2SO_4(aq)$	strong
(d)	$PbI_2(s)$	weak

67.

	Solution	Electrolyte	Aqueous Solution
(a)	$HF(aq)$	weak	$HF(aq)$
(b)	$HBr(aq)$	strong	$H^+(aq)$ and $Br^-(aq)$
(c)	$HNO_3(aq)$	strong	$H^+(aq)$ and $NO_3^-(aq)$
(d)	$HNO_2(aq)$	weak	$HNO_2(aq)$

69.

	Solution	Electrolyte	Aqueous Solution
(a)	$AgF(aq)$	strong	$Ag^+(aq)$ and $F^-(aq)$
(b)	$AgI(s)$	weak	$AgI(s)$
(c)	$Hg_2Cl_2(s)$	weak	$Hg_2Cl_2(s)$
(d)	$NiCl_2(aq)$	strong	$Ni^{2+}(aq)$ and $2\ Cl^-(aq)$

Section 14.11 *Net Ionic Equations*

71.
(1) Complete and balance the nonionized equation.
(2) Convert the nonionized equation into a total ionic equation.
(3) Cancel spectator ions to obtain the net ionic equation.
(4) Check (√) each ion or atom on both sides of the equation.

73. (a) Nonionized equation:

$$HCl(aq) + KOH(aq) \rightarrow KCl(aq) + H_2O(l)$$

Total ionic equation:

$$H^+(aq) + Cl^-(aq) + K^+(aq) + OH^-(aq) \rightarrow K^+(aq) + Cl^-(aq) + H_2O(l)$$

Net ionic equation:

$$H^+(aq) + OH^-(aq) \rightarrow H_2O(l)$$

(b) Nonionized equation:
$2\ HC_2H_3O_2(aq) + Ca(OH)_2(aq) \rightarrow Ca(C_2H_3O_2)_2(aq) + 2\ H_2O(l)$

Total ionic equation:
$2\ HC_2H_3O_2(aq) + Ca^{2+}(aq) + 2\ OH^-(aq) \rightarrow$
$Ca^{2+}(aq) + 2\ C_2H_3O_2^-(aq) + 2\ H_2O(l)$

Net ionic equation:
$HC_2H_3O_2(aq) + OH^-(aq) \rightarrow C_2H_3O_2^-(aq) + H_2O(l)$

75. (a) Nonionized equation:
$AgNO_3(aq) + KI(aq) \rightarrow AgI(s) + KNO_3(aq)$

Total ionic equation:
$Ag^+(aq) + NO_3^-(aq) + K^+(aq) + I^-(aq) \rightarrow AgI(s) + K^+(aq) + NO_3^-(aq)$

Net ionic equation:
$Ag^+(aq) + I^-(aq) \rightarrow AgI(s)$

(b) Nonionized equation:
$BaCl_2(aq) + K_2CrO_4(aq) \rightarrow BaCrO_4(s) + 2\ KCl(aq)$

Total ionic equation:
$Ba^{2+}(aq) + 2\ Cl^-(aq) + 2\ K^+(aq) + CrO_4^{2-}(aq) \rightarrow$
$BaCrO_4(s) + 2\ K^+(aq) + 2\ Cl^-(aq)$

Net ionic equation:
$Ba^{2+}(aq) + CrO_4^{2-}(aq) \rightarrow BaCrO_4(s)$

General Exercises

77. The color of methyl red is yellow at pH 7, while phenolphthalein is colorless. Therefore, the color of the water with both indicators is *yellow*.

79. The color of methyl orange is red at pH 3.2 and yellow at pH 4.4. Therefore, the color of a solution at pH 3.8 is *orange* (red + yellow).

81. Notice that NaH_2PO_4 is a *proton acceptor* in the first reaction and a *proton donor* in the second reaction. Thus, NaH_2PO_4 is *amphiprotic*.

83. $HCl(aq) \rightarrow H^+(aq) + Cl^-(aq)$
$0.50\ M\ HCl \rightarrow 0.50\ M\ H^+$

$$pH = -\log [H^+]$$
$$pH = -\log 0.50$$
$$pH = 0.30$$

85. $$1\ \cancel{\text{mL water}} \times \frac{3 \times 10^{22}\ \cancel{\text{molecules } H_2O}}{1\ \cancel{\text{mL water}}} \times \frac{2\ H^+}{1 \times 10^{9}\ \cancel{\text{molecules } H_2O}} = 6 \times 10^{13}\ H^+$$

87. Nonionized equation:
$NH_4OH(aq) + HC_2H_3O_2(aq) \rightarrow NH_4C_2H_3O_2(aq) + H_2O(l)$

Total ionic equation:
$NH_4OH(aq) + HC_2H_3O_2(aq) \rightarrow NH_4^+(aq) + C_2H_3O_2^-(aq) + H_2O(l)$

Net ionic equation:
$NH_4OH(aq) + HC_2H_3O_2(aq) \rightarrow NH_4^+(aq) + C_2H_3O_2^-(aq) + H_2O(l)$

Challenge Exercises

89. $HAla(aq) + NaOH(aq) \rightarrow NaAla(aq) + H_2O(l)$

1 mole HAla = 1 mole NaOH

$$21.05\ \cancel{\text{mL solution}} \times \frac{0.145\ \cancel{\text{mol NaOH}}}{1000\ \cancel{\text{mL solution}}} \times \frac{1\ \text{mol HAla}}{1\ \cancel{\text{mol NaOH}}} = 0.00305\ \text{mol HAla}$$

$$\text{MM of alanine} = \frac{0.272\ \text{g HAla}}{0.00305\ \text{mol HAla}} = 89.2\ \text{g/mol}$$

CHAPTER

Advanced Problem Solving 15

Section 15.1 *Advanced Problem-Solving Strategies*

1. (a) unit analysis (b) algebraic analysis

3. (a) meter (m) (b) gram (g)
(c) liter (L) (d) second (s)
(e) degree Celsius (°C) (f) calorie (cal)

5. (a) length (b) mass
(c) volume (d) time
(e) temperature (f) heat energy

7. (a) velocity (b) density
(c) specific heat (d) m/m % concentration

9. (a) $1 \text{ g Mg} \times \frac{1 \text{ mol Mg}}{20 \text{ g Mg}} \times \frac{1 \text{ mol MgO}}{1 \text{ mol Mg}} \times \frac{40 \text{ g MgO}}{1 \text{ mol MgO}} = \text{g MgO}$

Estimated answer: ~2 g MgO *Calculated answer: 2.036 g MgO*

(b) $1 \text{ g Ag} \times \frac{1 \text{ mol Ag}}{100 \text{ g Ag}} \times \frac{1 \text{ mol Ag}_2\text{O}}{2 \text{ mol Ag}} \times \frac{200 \text{ g Ag}_2\text{O}}{1 \text{ mol Ag}_2\text{O}} = \text{g Ag}_2\text{O}$

Estimated answer: ~1 g Ag_2O *Calculated answer: 1.235 g* Ag_2O

Section 15.2 *Concept Maps – Mole Relationships*

11. (a) molecules SO_2 → moles SO_2 → volume SO_2 (STP)

(b) volume N_2 (STP) → moles N_2 → mass N_2

13. (a) mass of reactant → moles of reactant → moles of product → mass of product

(b) mass of reactant → moles of reactant → moles of product → volume of product

15. (a) $0.0142 \cancel{\text{mol HCl}} \times \dfrac{36.46 \text{ g HCl}}{1 \cancel{\text{mol HCl}}} = 0.518 \text{ g HCl}$

(b) $0.0142 \cancel{\text{mol HCl}} \times \dfrac{22.4 \text{ L HCl}}{1 \cancel{\text{mol HCl}}} = 0.318 \text{ L HCl}$

(c) $0.0142 \cancel{\text{mol HCl}} \times \dfrac{6.02 \times 10^{23} \text{ molecules HCl}}{1 \cancel{\text{mol HCl}}} = 8.55 \times 10^{21} \text{ molecules HCl}$

(d) $\dfrac{0.0142 \text{ mol HCl}}{47.5 \cancel{\text{mL solution}}} \times \dfrac{1000 \cancel{\text{mL solution}}}{1 \text{ L solution}} = 0.299\ M \text{ HCl}$

17. (a) $1.00 \cancel{\text{g HF}} \times \dfrac{1 \cancel{\text{mol HF}}}{20.01 \cancel{\text{g HF}}} \times \dfrac{22.4 \text{ L HF}}{1 \cancel{\text{mol HF}}} = 1.12 \text{ L HF}$

(b) $1.00 \cancel{\text{g HF}} \times \dfrac{\cancel{\text{1mol HF}}}{20.01 \cancel{\text{g HF}}} \times \dfrac{6.02 \times 10^{23} \text{ molecules HF}}{1 \cancel{\text{mol HF}}}$

$= 3.01 \times 10^{22} \text{ molecules HF}$

(c) $\dfrac{1.00 \cancel{\text{g HF}}}{100.0 \cancel{\text{mL solution}}} \times \dfrac{1 \text{ mol HF}}{20.01 \cancel{\text{g HF}}} \times \dfrac{1000 \cancel{\text{mL solution}}}{1 \text{ L solution}} = 0.500\ M \text{ HF}$

19. (a) $1.00 \text{ L } CO_2 \times \frac{1 \text{ mol } CO_2}{22.4 \text{ L } CO_2} \times \frac{44.01 \text{ g } CO_2}{1 \text{ mol } CO_2} = 1.96 \text{ g } CO_2$

(b) $1.00 \text{ L } CO_2 \times \frac{1 \text{ mol } CO_2}{22.4 \text{ L } CO_2} \times \frac{6.02 \times 10^{23} \text{ molecules } CO_2}{1 \text{ mol } CO_2}$
$= 2.69 \times 10^{22} \text{ molecules } CO_2$

(c) $\frac{1.00 \text{ L } CO_2}{500.0 \text{ mL solution}} \times \frac{1 \text{ mol } CO_2}{22.4 \text{ L } CO_2} \times \frac{1000 \text{ mL solution}}{1 \text{ L solution}}$
$= 0.0893\ M\ H_2CO_3$

21. (a) $2.22 \times 10^{22} \text{ molecules } H_2S \times \frac{1 \text{ mol } H_2S}{6.02 \times 10^{23} \text{ molecules } H_2S} \times \frac{22.4 \text{ L } H_2S}{1 \text{ mol } H_2S}$
$= 0.826 \text{ L } H_2S$

(b) $2.22 \times 10^{22} \text{ molecules } H_2S \times \frac{1 \text{ mol } H_2S}{6.02 \times 10^{23} \text{ molecules } H_2S} \times \frac{34.1 \text{ g } H_2S}{1 \text{ mol } H_2S}$
$= 1.26 \text{ g } H_2S$

(c) $\frac{2.22 \times 10^{22} \text{ molecules } H_2S}{450.0 \text{ mL solution}} \times \frac{1 \text{ mol } H_2S}{6.02 \times 10^{23} \text{ molecules } H_2S} \times \frac{1000 \text{ mL solution}}{1 \text{ L solution}}$
$= 0.0819\ M\ H_2S$

Section 15.3 *Concept Maps – Stoichiometry*

23. $CH_4(g) + 2\,O_2(g) \rightarrow CO_2(g) + 2\,H_2O(l)$

(a) $5.00 \text{ g } H_2O \times \frac{1 \text{ mol } H_2O}{18.02 \text{ g } H_2O} \times \frac{1 \text{ mol } CH_4}{2 \text{ mol } H_2O} \times \frac{16.04 \text{ g } CH_4}{1 \text{ mol } CH_4} = 2.23 \text{ g } CH_4$

(b) $5.00 \text{ g } H_2O \times \frac{1 \text{ mol } H_2O}{18.02 \text{ g } H_2O} \times \frac{1 \text{ mol } CO_2}{2 \text{ mol } H_2O} \times \frac{22.4 \text{ L } CO_2}{1 \text{ mol } CO_2} = 3.11 \text{ L } CO_2$

25. $2\,Al(s) + 6\,HCl(aq) \rightarrow 2\,AlCl_3(aq) + 3\,H_2(g)$

(a) $0.466 \text{ g Al} \times \frac{1 \text{ mol Al}}{26.98 \text{ g Al}} \times \frac{3 \text{ mol } H_2}{2 \text{ mol Al}} \times \frac{22.4 \text{ L } H_2}{1 \text{ mol } H_2} = 0.580 \text{ L } H_2$

(b) $0.466 \text{ g Al} \times \frac{1 \text{ mol Al}}{26.98 \text{ g Al}} \times \frac{6 \text{ mol HCl}}{2 \text{ mol Al}} \times \frac{1 \text{ L solution}}{0.100 \text{ mol HCl}} \times \frac{1000 \text{ mL solution}}{1 \text{ L solution}}$
$= 518 \text{ mL HCl solution}$

27. $MgBr_2(aq) + 2\ AgNO_3(aq) \rightarrow 2\ AgBr(s) + Mg(NO_3)_2(aq)$

(a) $$50.0\ \cancel{mL\ MgBr_2} \times \frac{0.100\ \cancel{mol\ MgBr_2}}{1000\ \cancel{mL\ MgBr_2}} \times \frac{2\ mol\ AgNO_3}{1\ \cancel{mol\ MgBr_2}} = 0.0100\ mol\ AgNO_3$$

$$\frac{0.0100\ mol\ AgNO_3}{13.9\ \cancel{mL\ solution}} \times \frac{1000\ \cancel{mL\ solution}}{1\ L\ solution} = 0.719\ M\ AgNO_3$$

(b) $$50.0\ \cancel{mL\ MgBr_2} \times \frac{0.100\ \cancel{mol\ MgB_2}}{1000\ \cancel{mL\ MgB_2}} \times \frac{2\ mol\ AgBr}{1\ \cancel{mol\ MgBr_2}} \times \frac{187.77\ g\ AgBr}{1\ mol\ AgBr}$$

$$= 1.88\ g\ AgBr$$

Section 15.4 *Multiple-Reaction Stoichiometry*

29. $2\ C(s) + O_2(g) \rightarrow 2\ CO(g)$

$2\ CO(g) + O_2(g) \rightarrow 2\ CO_2(g)$

(a) $$25.0\ \cancel{g\ C} \times \frac{1\ \cancel{mol\ C}}{12.01\ \cancel{g\ C}} \times \frac{2\ \cancel{mol\ CO}}{2\ \cancel{mol\ C}} \times \frac{2\ \cancel{mol\ CO_2}}{2\ \cancel{mol\ CO}} \times \frac{44.01\ g\ CO_2}{1\ \cancel{mol\ CO_2}} = 91.6\ g\ CO_2$$

(b) $$25.0\ \cancel{g\ C} \times \frac{1\ \cancel{mol\ C}}{12.01\ \cancel{g\ C}} \times \frac{2\ \cancel{mol\ CO}}{2\ \cancel{mol\ C}} \times \frac{2\ \cancel{mol\ CO_2}}{2\ \cancel{mol\ CO}} \times \frac{22.4\ L\ CO_2}{1\ \cancel{mol\ CO_2}} = 46.6\ L\ CO_2$$

(c) 1st reaction

$$25.0\ \cancel{g\ C} \times \frac{1\ \cancel{mol\ C}}{12.01\ \cancel{g\ C}} \times \frac{1\ \cancel{mol\ O_2}}{2\ \cancel{mol\ C}} \times \frac{32.00\ g\ O_2}{1\ \cancel{mol\ O_2}} = 33.3\ g\ O_2$$

2nd reaction

$$25.0\ \cancel{g\ C} \times \frac{1\ \cancel{mol\ C}}{12.01\ \cancel{g\ C}} \times \frac{2\ \cancel{mol\ CO}}{2\ \cancel{mol\ C}} \times \frac{1\ \cancel{mol\ O_2}}{2\ \cancel{mol\ CO}} \times \frac{32.00\ g\ O_2}{1\ \cancel{mol\ O_2}} = 33.3\ g\ O_2$$

Total mass of O_2 consumed: $33.3\ g + 33.3\ g = 66.6\ g\ O_2$

$S(s) + O_2(g) \rightarrow SO_2(g)$

$2\,SO_2(g) + O_2(g) \rightarrow 2\,SO_3(g)$

$SO_3(g) + H_2O(l) \rightarrow H_2SO_4(l)$

(a) $$1.00\ \cancel{\text{kg S}} \times \frac{1000\ \cancel{\text{g S}}}{1\ \cancel{\text{kg S}}} \times \frac{1\ \cancel{\text{mol S}}}{32.07\ \cancel{\text{g S}}} \times \frac{1\ \cancel{\text{mol } SO_2}}{1\ \cancel{\text{mol S}}} \times \frac{2\ \cancel{\text{mol } SO_3}}{2\ \cancel{\text{mol } SO_2}} \times \frac{80.07\ \text{g } SO_3}{1\ \cancel{\text{mol } SO_3}} = 2{,}500\ \text{g } SO_3$$

(b) $$1.00\ \cancel{\text{kg S}} \times \frac{1000\ \cancel{\text{g S}}}{1\ \cancel{\text{kg S}}} \times \frac{1\ \cancel{\text{mol S}}}{32.07\ \cancel{\text{g S}}} \times \frac{1\ \cancel{\text{mol } SO_2}}{1\ \cancel{\text{mol S}}} \times \frac{2\ \cancel{\text{mol } SO_3}}{2\ \cancel{\text{mol } SO_2}} \times \frac{22.4\ \text{L } SO_3}{1\ \cancel{\text{mol } SO_3}} = 698\ \text{L } SO_3$$

(c) $$1.00\ \cancel{\text{kg S}} \times \frac{1000\ \cancel{\text{g S}}}{1\ \cancel{\text{kg S}}} \times \frac{1\ \cancel{\text{mol S}}}{32.07\ \cancel{\text{g S}}} \times \frac{1\ \cancel{\text{mol } SO_2}}{1\ \cancel{\text{mol S}}} \times \frac{2\ \cancel{\text{mol } SO_3}}{2\ \cancel{\text{mol } SO_2}} \times \frac{1\ \cancel{\text{mol } H_2SO_4}}{1\ \cancel{\text{mol } SO_3}}$$

$$\times \frac{98.09\ \text{g } H_2SO_4}{1\ \cancel{\text{mol } H_2SO_4}} \times 55.0\% = 1{,}680\ \text{g } H_2SO_4$$

33. $2\,NH_3(aq) + NaOCl(aq) \rightarrow N_2H_4(l) + NaCl(aq) + H_2O(l)$

$2\,N_2H_4(l) + N_2O_4(l) \rightarrow 3\,N_2(g) + 4\,H_2O(g)$

(a) $$50.0\ \cancel{\text{mL } NH_3} \times \frac{6.00\ \cancel{\text{mol } NH_3}}{1000\ \cancel{\text{mL } NH_3}} \times \frac{1\ \cancel{\text{mol } N_2H_4}}{2\ \cancel{\text{mol } NH_3}} \times \frac{3\ \cancel{\text{mol } N_2}}{2\ \cancel{\text{mol } N_2H_4}} \times \frac{28.02\ \text{g } N_2}{1\ \cancel{\text{mol } N_2}} = 6.30\ \text{g } N_2$$

(b) $$50.0\ \cancel{\text{mL } NH_3} \times \frac{6.00\ \cancel{\text{mol } NH_3}}{1000\ \cancel{\text{mL } NH_3}} \times \frac{1\ \cancel{\text{mol } N_2H_4}}{2\ \cancel{\text{mol } NH_3}} \times \frac{3\ \cancel{\text{mol } N_2}}{2\ \cancel{\text{mol } N_2H_4}} \times \frac{22.4\ \text{L } N_2}{1\ \cancel{\text{mol } N_2}} = 5.04\ \text{L } N_2$$

(c) $$50.0\ \cancel{\text{mL } NH_3} \times \frac{6.00\ \cancel{\text{mol } NH_3}}{1000\ \cancel{\text{mL } NH_3}} \times \frac{1\ \cancel{\text{mol } N_2H_4}}{2\ \cancel{\text{mol } NH_3}} \times \frac{4\ \cancel{\text{mol } H_2O}}{2\ \cancel{\text{mol } N_2H_4}} \times \frac{18.02\ \text{g } H_2O}{1\ \cancel{\text{mol } H_2O}} = 5.41\ \text{g } H_2O$$

Section 15.5 *Advanced Problem-Solving Examples*

35. $PV = nRT;\ n = g/MM$

thus, $MM = \dfrac{g\,R\,T}{P\,V}$

$P = 0.989 \text{ atm};\ 1.45 \text{ g} = 1 \text{ L};\ T = 100\ °\text{C} + 273 = 373 \text{ K};\ R = \dfrac{0.0821 \text{ atm} \cdot \text{L}}{1 \text{ mol} \cdot \text{K}}$

$$MM = \frac{1.45 \text{ g} \times 373\ \cancel{\text{K}}}{0.989\ \cancel{\text{atm}} \times 1\ \cancel{\text{L}}} \times \frac{0.0821\ \cancel{\text{atm}} \cdot \cancel{\text{L}}}{1 \text{ mol} \cdot \cancel{\text{K}}} = 44.9 \text{ g/mol}$$

37. $CaCl_2(aq) \rightarrow Ca^{2+}(aq) + 2\,Cl^-(aq)$

$$34.5\ \cancel{\text{g CaCl}_2} \times \frac{1\ \cancel{\text{mol CaCl}_2}}{110.98\ \cancel{\text{g CaCl}_2}} \times \frac{1 \text{ mol Ca}^{2+}}{1\ \cancel{\text{mol CaCl}_2}} = 0.311 \text{ mol Ca}^{2+}$$

$$34.5\ \cancel{\text{g CaCl}_2} \times \frac{1\ \cancel{\text{mol CaCl}_2}}{110.98\ \cancel{\text{g CaCl}_2}} \times \frac{2 \text{ mol Cl}^-}{1\ \cancel{\text{mol CaCl}_2}} = 0.622 \text{ mol Cl}^-$$

Molarity of Ca^{2+}: $\dfrac{0.311 \text{ mol Ca}^{2+}}{500.0\ \cancel{\text{mL solution}}} \times \dfrac{1000\ \cancel{\text{mL solution}}}{1 \text{ L solution}} = 0.622\ M \text{ Ca}^{2+}$

Molarity of Cl^-: $\dfrac{0.622 \text{ mol Cl}^-}{500.0\ \cancel{\text{mL solution}}} \times \dfrac{1000\ \cancel{\text{mL solution}}}{1 \text{ L solution}} = 1.24\ M \text{ Cl}^-$

39. $Zn(s) + 2\,HCl(aq) \rightarrow ZnCl_2(aq) + H_2(g)$

$$50.0\ \cancel{\text{mL H}_2} \times \frac{1\ \cancel{\text{L H}_2}}{1000\ \cancel{\text{mL H}_2}} \times \frac{1\ \cancel{\text{mol H}_2}}{22.4\ \cancel{\text{L H}_2}} \times \frac{2\ \cancel{\text{mol HCl}}}{1\ \cancel{\text{mol H}_2}} \times \frac{1000 \text{ mL solution}}{0.100\ \cancel{\text{mol HCl}}}$$

$$= 44.6 \text{ mL solution}$$

41. $PV = nRT; \quad n = \dfrac{PV}{RT}$

$$P = 749 \cancel{\text{mm Hg}} \times \frac{1.00 \text{ atm}}{760 \cancel{\text{mm Hg}}} = 0.986 \text{ atm}$$

$$V = 1.50 \text{ L}; \; T = 25°\text{C} + 273 = 298 \text{ K}; \; R = \frac{0.0821 \text{ atm} \cdot \text{L}}{1 \text{ mol} \cdot \text{K}}$$

$$n = \frac{0.986 \cancel{\text{atm}} \times 0.150 \cancel{\text{L}}}{298 \cancel{\text{K}}} \times \frac{1 \text{ mol} \cdot \cancel{\text{K}}}{0.0821 \cancel{\text{atm}} \cdot \cancel{\text{L}}} = 0.00605 \text{ mol } N_2O$$

$$0.00605 \cancel{\text{mol } N_2O} \times \frac{6.02 \times 10^{23} \text{ molecules } N_2O}{1 \cancel{\text{mol } N_2O}} = 3.64 \times 10^{21} \text{ molecules } N_2O$$

$$0.00605 \cancel{\text{mol } N_2O} \times \frac{44.02 \text{ g } N_2O}{1 \cancel{\text{mol } N_2O}} = 0.266 \text{ g } N_2O$$

43. $39.6 \cancel{\text{g C}} \times \dfrac{1 \text{ mol C}}{12.01 \text{ g C}} = 3.30 \text{ mol C}$

$$7.7 \cancel{\text{g H}} \times \frac{1 \text{ mol H}}{1.01 \text{ g H}} = 7.6 \text{ mol H}$$

$$52.8 \cancel{\text{g O}} \times \frac{1 \text{ mol O}}{16.00 \text{ g O}} = 3.30 \text{ mol O}$$

$$\frac{C_{3.30}}{3.30}\frac{H_{7.6}}{3.30}\frac{O_{3.30}}{3.30} = C_{1.00}H_{2.3}O_{1.00}$$

Empirical Formula: $3\,(C_{1.00}H_{2.3}O_{1.00}) = C_3H_7O_3$

MM of Empirical Formula: $3(12.01) \text{ g} + 7(1.01) \text{ g} + 3(16.00) \text{ g} = 91.10 \text{ g/mol}$

$$\text{MM: } \frac{1 \cancel{\text{mL}}}{4.93 \times 10^{21} \cancel{\text{molecules}}} \times \frac{1.47 \text{ g}}{1 \cancel{\text{mL}}} \times \frac{6.02 \times 10^{23} \cancel{\text{molecules}}}{1 \text{ mol}} = 180 \text{ g/mol}$$

$$\text{Molecular Formula: } \frac{180 \text{ g/mol}}{91.0 \text{ g/mol}} \approx 2$$

Molecular Formula: $2\,(C_3H_7O_3) = C_6H_{14}O_6$

45. $2\ NaClO_3(s) \xrightarrow[\Delta]{MnO_2} 2\ NaCl(s) + 3\ O_2(g)$

$P_{total} = P_{oxygen} + P_{water\ vapor}$

$P_{oxygen} = 754\text{ mm Hg} - 24\text{ mm Hg} = 730\text{ mm Hg}$

$$P = 730\text{ mm Hg} \times \frac{1.00\text{ atm}}{760\text{ mm Hg}} = 0.961\text{ atm}$$

$$V = 245\text{ mL} = 0.245\text{ L};\ T = 25°\text{C} + 273 = 298\text{ K};\ R = \frac{0.0821\text{ atm}\cdot\text{L}}{1\text{ mol}\cdot\text{K}}$$

$$PV = nRT;\ n = \frac{P\,V}{R\,T}$$

$$n = \frac{0.961\text{ atm} \times 0.245\text{ L}}{298\text{ K}} \times \frac{1\text{ mol}\cdot\text{K}}{0.0821\text{ atm}\cdot\text{L}} = 0.00962\text{ mol } O_2$$

$$0.00962\text{ mol } O_2 \times \frac{2\text{ mol } NaClO_3}{3\text{ mol } O_2} \times \frac{106.44\text{ g } NaClO_3}{1\text{ mol } NaClO_3} = 0.683\text{ g } NaClO_3$$

General Exercises

47. $$288{,}125\text{ pencils} \times \frac{1\text{ gross}}{144\text{ pencils}} = 2000.87\text{ gross}$$

Rounded to a whole number, the number is 2000 gross.

49. $N_2(g) + 2\ O_2(g) \rightarrow 2\ NO_2(g)$

$$45.0\text{ mL } N_2 \times \frac{2\text{ mL } NO_2}{1\text{ mL } N_2} = 90.0\text{ mL } NO_2$$

$$65.0\text{ mL } O_2 \times \frac{2\text{ mL } NO_2}{2\text{ mL } O_2} = 65.0\text{ mL } NO_2$$

(a) O_2 is the limiting reactant
(b) 65.0 mL NO_2

51. $2\ Al(OH)_3(s) + 3\ H_2SO_4(aq) \rightarrow Al_2(SO_4)_3(aq) + 6\ H_2O(l)$

$$1.00\ \cancel{g\ Al(OH)_3} \times \frac{1\ \cancel{mol\ Al(OH)_3}}{78.01\ \cancel{g\ Al(OH)_3}} \times \frac{6\ \cancel{mol\ H_2O}}{2\ \cancel{mol\ Al(OH)_3}} \times \frac{18.02\ g\ H_2O}{1\ \cancel{mol\ H_2O}} = 0.693\ g\ H_2O$$

$$25.0\ \cancel{mL\ H_2SO_4} \times \frac{0.500\ \cancel{mol\ H_2SO_4}}{1000\ \cancel{mL\ H_2SO_4}} \times \frac{6\ \cancel{mol\ H_2O}}{3\ \cancel{mol\ H_2SO_4}} \times \frac{18.02\ g\ H_2O}{1\ \cancel{mol\ H_2O}} = 0.451\ g\ H_2O$$

(a) H_2SO_4 is the limiting reactant
(b) 0.451 g H_2O

53. $$12.0\ \cancel{g\ Au} \times \frac{1\ \cancel{mol\ Au}}{196.97\ \cancel{g\ Au}} \times \frac{6.02 \times 10^{23}\ \cancel{atoms\ Au}}{1\ \cancel{mol\ Au}} \times \frac{1\ \cancel{mL\ seawater}}{1.22 \times 10^{10}\ \cancel{atoms\ Au}} \times$$

$$\frac{1.05\ \cancel{g\ seawater}}{1\ \cancel{mL\ seawater}} \times \frac{1\ kg\ seawater}{1000\ \cancel{g\ seawater}} = 3.16 \times 10^{9}\ kg\ seawater$$

55. $$50.0\ \cancel{mL\ NaCl\ solution} \times \frac{0.100\ \cancel{mol\ NaCl}}{1000\ \cancel{mL\ NaCl\ solution}} \times \frac{1\ mol\ Na^+}{1\ \cancel{mol\ NaCl}}$$
$$= 0.00500\ mol\ Na^+\ from\ NaCl$$

$$50.0\ \cancel{mL\ Na_2SO_4\ solution} \times \frac{0.200\ \cancel{mol\ Na_2SO_4}}{1000\ \cancel{mL\ Na_2SO_4}\ solution} \times \frac{2\ mol\ Na^+}{1\ \cancel{mol\ Na_2SO_4}}$$
$$= 0.0200\ mol\ Na^+\ from\ Na_2SO_4$$

Total mol Na^+: 0.00500 mol + 0.0200 mol = 0.0250 mol Na^+

$$\frac{0.0250\ mol\ Na^+}{(50.0 + 50.0)\ \cancel{mL\ solution}} \times \frac{1000\ \cancel{mL\ solution}}{1\ L\ solution} = 0.250\ M\ Na^+$$

57. $Na_2CO_3(s) + 2\ HCl(aq) \rightarrow 2\ NaCl(aq) + H_2O(l) + CO_2(g)$
$HCl(aq) + NaOH(aq) \rightarrow NaCl(aq) + H_2O(l)$

Total HCl: $$50.0\ \cancel{mL\ HCl} \times \frac{0.345\ mol\ HCl}{1000\ \cancel{mL\ HCl}} = 0.0173\ mol\ HCl$$

Excess HCl:
$$15.9\ \cancel{mL\ NaOH} \times \frac{0.155\ \cancel{mol\ NaOH}}{1000\ \cancel{mL\ NaOH}} \times \frac{1\ mol\ HCl}{1\ \cancel{mol\ NaOH}} = 0.00246\ mol\ HCl$$

HCl neutralized: (0.0173 – 0.00246) mol HCl = 0.0148 mol HCl

$$0.0148\ \cancel{mol\ HCl} \times \frac{1\ \cancel{mol\ Na_2CO_3}}{2\ \cancel{mol\ HCl}} \times \frac{105.99\ g\ Na_2CO_3}{1\ \cancel{mol\ Na_2CO_3}} = 0.784\ g\ Na_2CO_3$$

Challenge Exercises

59. A concept map for the college administration is:

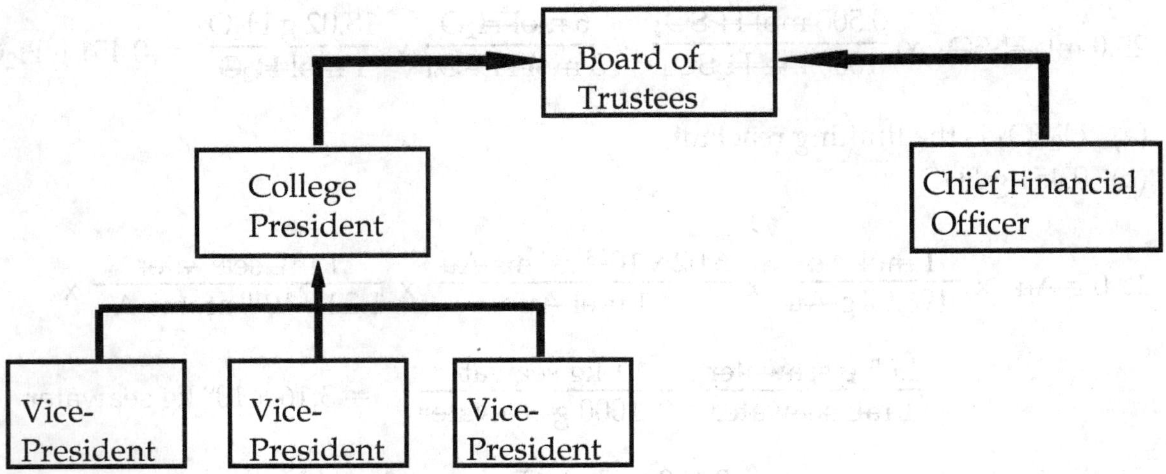

CHAPTER

16

Chemical Equilibrium

Section 16.1 *Collision Theory*

1. Theoretical Factors that Decrease the Rate of a Reaction
   (a) Decreasing collision frequency of molecules *decreases* the rate.
   (b) Decreasing collision energy of molecules *decreases* the rate.
   (c) Ineffective collision orientation of molecules *decreases* the rate.

3. Effective Collision Orientation

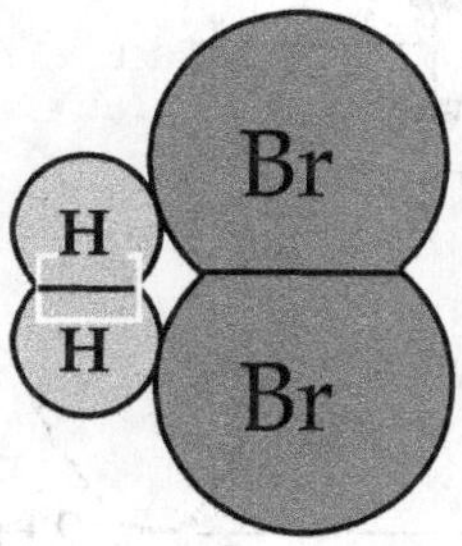

5.

	Reaction Conditions	Effect on the Rate of Reaction
(a)	decrease the concentration of a reactant	The rate of reaction *decreases* because fewer collisions occur.
(b)	decrease the temperature of the reaction	The rate of reaction *decreases* because both collision frequency and collision energy decrease.
(c)	add an acid catalyst	The rate of reaction *increases* because it speeds up the reaction.

7. No, the catalyst speeds up the rate at which sulfur trioxide is produced, but does not increase the amount of sulfur trioxide produced.

Section 16.2 *Energy Profiles of Chemical Reactions*

9. $PCl_5(g)$ + *heat* $\rightleftarrows$ $PCl_3(g)$ + $Cl_2(g)$

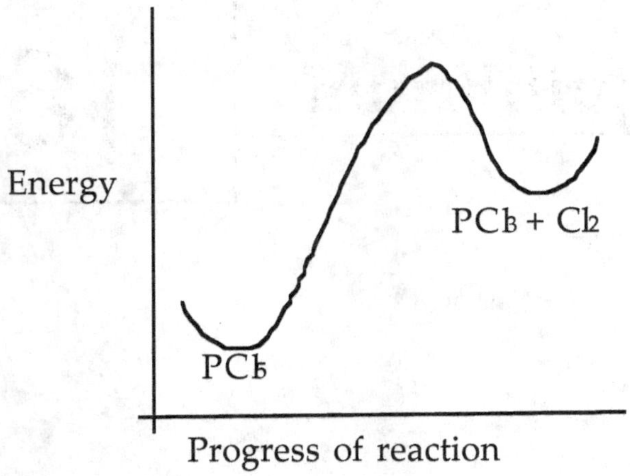

11. $H_2(g)$ + $Cl_2(g)$ $\rightleftarrows$ 2 $HCl(g)$ + *heat*

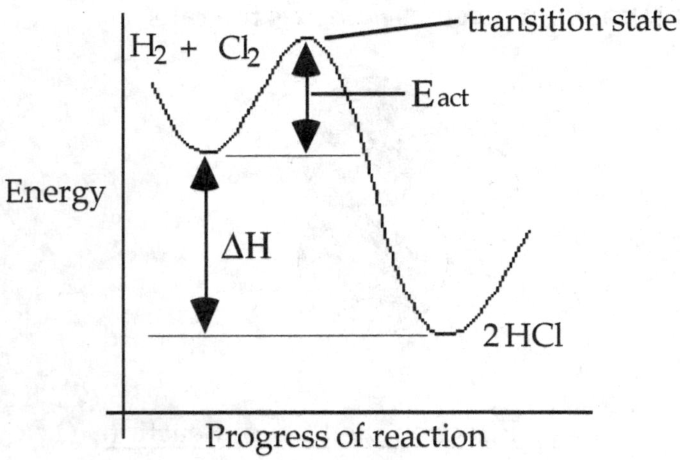

13. A catalyst is a substance that allows a reaction to proceed faster by lowering the energy of activation (E_{act}).

Section 16.3 *The Chemical Equilibrium Concept*

15. (a) **False**. In the general equilibrium expression, the K_{eq} is dependent on temperature and its value varies at different temperatures.

(b) **True**. The amount of a solid does not affect a gaseous equilibrium, and thus is omitted from the K_{eq} expression. A substance participating in a different physical state (solid or liquid) is assigned a concentration of [1], and does not appear in the K_{eq} expression.

Section 16.4 *General Equilibrium Constant, K_{eq}*

17.

	Equilibrium Reaction	Equilibrium Constant Expression
(a)	$H_2(g) + F_2(g) \rightleftarrows 2\,HF(g)$	$K_{eq} = \dfrac{[HF]^2}{[H_2]\,[F_2]}$
(b)	$4\,NH_3(g) + 7\,O_2(g) \rightleftarrows 4\,NO_2(g) + 6\,H_2O(g)$	$K_{eq} = \dfrac{[NO_2]^4\,[H_2O]^6}{[NH_3]^4\,[O_2]^7}$
(c)	$ZnCO_3(s) \rightleftarrows ZnO(s) + CO_2(g)$	$K_{eq} = [CO_2]$

19. $2\,SO_2(g) + O_2(g) \rightleftarrows 2\,SO_3(g)$

$[SO_2] = 1.75\,M \qquad [O_2] = 1.50\,M \qquad [SO_3] = 2.25\,M$

$$K_{eq} = \frac{[SO_3]^2}{[SO_2]^2\,[O_2]} = \frac{(2.25)^2}{(1.75)^2\,(1.50)} = 1.10$$

Section 16.5 *Equilibria Shifts for Gases*

21. $N_2(g) + 3\,H_2(g) \rightleftarrows 2\,NH_3(g) + \textit{heat}$

(a) If the temperature increases, the heat shifts the equilibrium to the left. Thus, the concentration of NH_3 *decreases*.

(b) If the pressure increases, the pressure shifts the equilibrium to the right, which has fewer molecules. Thus, the concentration of NH_3 *increases*.

23. $CH_4(g) + H_2O(g) + \textit{heat} \rightleftarrows CO(g) + 3\,H_2(g)$

(a)	increase $[CH_4]$	shifts right	(b)	decrease $[H_2O]$	shifts left
(c)	increase [CO]	shifts left	(d)	decrease $[H_2]$	shifts right
(e)	increase temp.	shifts right	(f)	decrease temp.	shifts left
(g)	increase volume	no shift	(h)	decrease volume	no shift
(i)	add Zn dust	no shift	(j)	add Xe inert gas	no shift

Section 16.6 *Ionization Equilibrium Constant, K_i*

25. **Equilibrium Reaction** / **Equilibrium Constant Expression**

(a) $HCHO_2(aq) \rightleftarrows H^+(aq) + CHO_2^-(aq)$ $\quad K_i = \frac{[H^+]\,[CHO_2^-]}{[HCHO_2]}$

(b) $H_2C_2O_4(aq) \rightleftarrows H^+(aq) + HC_2O_4^-(aq)$ $\quad K_i = \frac{[H^+]\,[HC_2O_4^-]}{[H_2C_2O_4]}$

(c) $H_3C_6H_5O_7(aq) \rightleftarrows H^+(aq) + H_2C_6H_5O_7^-(aq)$ $\quad K_i = \frac{[H^+]\,[H_2C_6H_5O_7^-]}{[H_3C_6H_5O_7]}$

27. $HNO_2(aq) \rightleftarrows H^+(aq) + NO_2^-(aq)$

$[H^+] = [NO_2^-] = 7.5 \times 10^{-3}\,M \qquad [HNO_2] = 0.125\,M$

$$K_i = \frac{[H^+]\,[NO_2^-]}{[HNO_2]} = \frac{(7.5 \times 10^{-3})\,(7.5 \times 10^{-3})}{(0.125)} = 4.5 \times 10^{-4}$$

29. $HF(aq) \rightleftarrows H^+(aq) + F^-(aq)$

$[H^+] = 10^{-pH} = 10^{-2.00} = 0.010\,M$

$[H^+] = [F^-] = 0.010\,M \qquad [HF] = 0.139\,M$

$$K_i = \frac{[H^+]\,[F^-]}{[HF]} = \frac{(0.010)\,(0.010)}{(0.139)} = 7.2 \times 10^{-4}$$

Section 16.7 *Equilibria Shifts for Weak Acids and Bases*

31. $HF(aq) \rightleftarrows H^+(aq) + F^-(aq)$

(a) increase [HF] shift right (b) increase $[H^+]$ shift left
(c) decrease [HF] shift left (d) decrease $[F^-]$ shift right
(e) add solid NaF shift left (f) add gaseous HCl shift left
(g) add solid NaOH shift right (h) increase pH shift right

33. $HC_2H_3O_2(aq) \rightleftarrows H^+(aq) + C_2H_3O_2^-(aq)$

(a)	increase $[HC_2H_3O_2]$	shift right	(b)	increase $[H^+]$	shift left
(c)	decrease $[HC_2H_3O_2]$	shift left	(d)	decrease $[C_2H_3O_2^-]$	shift right
(e)	add solid $NaC_2H_3O_2$	shift left	(f)	add solid NaCl	no shift
(g)	add NaOH solid	shift right	(h)	increase pH	shift right

Section 16.8 *Solubility Product Equilibrium Constant, K_{sp}*

35.

	Reaction	Solubility Product Expression
(a)	$AgI(s) \rightleftarrows Ag^+(aq) + I^-(aq)$	$K_{sp} = [Ag^+][I^-]$
(b)	$Ag_2CrO_4(s) \rightleftarrows 2\,Ag^+(aq) + CrO_4^{2-}(aq)$	$K_{sp} = [Ag^+]^2[CrO_4^{2-}]$
(c)	$Ag_3PO_4(s) \rightleftarrows 3\,Ag^+(aq) + PO_4^{3-}(aq)$	$K_{sp} = [Ag^+]^3[PO_4^{3-}]$

37. $CoS(s) \rightleftarrows Co^{2+}(aq) + S^{2-}(aq)$

Because one mole of S^{2-} is produced for every mole of Co^{2+},

$[Co^{2+}] = [S^{2-}] = 7.7 \times 10^{-11}$

$K_{sp} = [Co^{2+}][S^{2-}] = (7.7 \times 10^{-11})(7.7 \times 10^{-11}) = 5.9 \times 10^{-21}$

Section 16.9 *Equilibria Shifts for Slightly Soluble Compounds*

39. $Ca_3(PO_4)_2(s) \rightleftarrows 3\,Ca^{2+}(aq) + 2\,PO_4^{3-}(aq)$

(a)	increase $[Ca^{2+}]$	shift left	(b)	increase $[PO_4^{3-}]$	shift left
(c)	decrease $[Ca^{2+}]$	shift right	(d)	decrease $[PO_4^{3-}]$	shift right
(e)	add solid $Ca_3(PO_4)_2$	no shift	(f)	add solid $Ca(NO_3)$	shift left
(g)	add solid KNO_3	no shift	(h)	decrease pH	shift right

41. $Cu(OH)_2(s) \rightleftarrows Cu^{2+}(aq) + 2\,OH^-(aq)$

(a)	increase $[Cu^{2+}]$	shift left	(b)	increase $[OH^-]$	shift left
(c)	decrease $[Cu^{2+}]$	shift right	(d)	decrease $[OH^-]$	shift right
(e)	add solid $Cu(OH)_2$	no shift	(f)	add solid NaOH	shift left
(g)	add solid NaCl	no shift	(h)	decrease pH	shift right

General Exercises

43. $N_2O_4(g) \rightleftarrows 2\,NO_2(g)$

$[N_2O_4] = 4.5 \times 10^{-5}\,M \qquad [NO_2] = 3.0 \times 10^{-3}\,M$

$$K_{eq} = \frac{[NO_2]^2}{[N_2O_4]} = \frac{(3.0 \times 10^{-3})^2}{(4.5 \times 10^{-5})} = 0.20$$

45. $H_2O(l) \rightleftarrows H^+(aq) + OH^-(aq)$

(a) increase $[H^+]$ shift left (b) decrease $[OH^-]$ shift right
(c) increase pH shift right (d) decrease pH shift left

Challenge Exercises

47. $Ca(OH)_2(s) \rightleftarrows Ca^{2+}(aq) + 2\,OH^-(aq)$

pH = 12.35

$[H^+] = 4.5 \times 10^{-13}\,M$

$$[OH^-] = \frac{1.0 \times 10^{-14}}{[H^+]} = \frac{1.0 \times 10^{-14}}{4.5 \times 10^{-13}} = 2.2 \times 10^{-2}\,M$$

Because two moles of OH^- are produced for every mole of Ca^{2+},

$$[Ca^{2+}] = \frac{1}{2}[OH^-] = \frac{1}{2}(2.2 \times 10^{-2}) = 1.1 \times 10^{-29}\,M$$

therefore,

$$K_{sp} = [Ca^{2+}][OH^-]^2 = (1.1 \times 10^{-2})(2.2 \times 10^{-2})^2 = 5.3 \times 10^{-6}$$

CHAPTER

17

Oxidation and Reduction

Section 17.1 *Oxidation Numbers*

1.

	Element	Ox. No.		Element	Ox. No.
(a)	Ag	0	(b)	Mg	0
(c)	S	0	(d)	F_2	0

Note: All elements in the free state have an oxidation number of zero.

3.

	Ion	Ox. No.		Ion	Ox. No.
(a)	Ag^+	+1	(b)	Mg^{2+}	+2
(c)	S^{2-}	-2	(d)	F^-	-1

5.

	Compound	Oxidation Number of Nitrogen
(a)	NH_3	ox. no. N + 3(ox. no. H) = 0 ox. no. N + 3(+1) = 0 ox. no. N + 3 = 0 ox. no. N = −3
(b)	N_2O_4	2(ox. no. N) + 4(ox. no. O) = 0 2(ox. no. N) + 4(−2) = 0 2(ox. no. N) + (−8) = 0 2(ox. no. N) = +8 ox. no. N = +4
(c)	Li_3N	3(ox. no. Li) + ox. no. N = 0 3(+1) + ox. no. N = 0 +3 + ox. no. N = 0 ox. no. N = −3
(d)	KNO_3	ox. no. K + ox. no. N + 3(ox. no. O) = 0 +1 + ox. no. N + 3(−2) = 0 +1 + ox. no. N + (−6) = 0 ox. no. N = +5

7. Ion — Oxidation Number of Sulfur

(a) SO_3^{2-}

$\text{ox. no. S} + 3(\text{ox. no. O}) = -2$
$\text{ox. no. S} + 3(-2) = -2$
$\text{ox. no. S} + (-6) = -2$
$\text{ox. no. S} = +4$

(b) HSO_4^-

$\text{ox. no. H} + \text{ox. no. S} + 4(\text{ox. no. O}) = -1$
$(+1) + \text{ox. no. S} + 4(-2) = -1$
$(+1) + \text{ox. no. S} + (-8) = -1$
$\text{ox. no. S} = +6$

(c) HS^-

$\text{ox. no. H} + \text{ox. no. S} = -1$
$(+1) + \text{ox. no. S} = -1$
$\text{ox. no. S} = -2$

(d) $S_2O_8^{2-}$

$2(\text{ox. no. S}) + 8(\text{ox. no. O}) = -2$
$2(\text{ox. no. S}) + 8(-2) = -2$
$2(\text{ox. no. S}) + (-16) = -2$
$2(\text{ox. no. S}) = +14$
$\text{ox. no. S} = +7$

Section 17.2 *Oxidation–Reduction Reactions*

9. (a) A redox process characterized by electron gain is termed *reduction.*
   (b) A substance undergoing reduction in a redox reaction is an *oxidizing agent*.

11. (a) Oxidized: Co — Reduced: S
    (b) Oxidized: Cd — Reduced: Cl_2

13. (a)

Al: 0 → +3; Cr: +3 → 0

$Al_{(s)} + Cr^{3+}{}_{(aq)} \rightarrow Al^{3+}{}_{(aq)} + Cr_{(s)}$

Oxidized: Al — Reduced: Cr^{3+}

(b)

F: 0 → −1; Cl: −1 → 0

$F_{2(g)} + 2\,Cl^-{}_{(aq)} \rightarrow 2\,F^-{}_{(aq)} + Cl_{2(g)}$

Oxidized: Cl^- — Reduced: F_2

15. (a)

0 ⟶ +1 (H)
+2 ⟶ 0 (Cu)

$CuO_{(s)} + H_{2(g)} \rightarrow Cu_{(s)} + H_2O_{(l)}$

Oxidizing agent: CuO Reducing agent: H_2

(b)

+2 ⟶ +4 (C)
+2 ⟶ 0 (Pb)

$PbO_{(s)} + CO_{(g)} \rightarrow Pb_{(s)} + CO_{2(g)}$

Oxidizing agent: PbO Reducing agent: CO

Section 17.3 *Balancing Redox Equations: Oxidation Number Method*

17. Yes, the total electron loss by the oxidation process always equals the total electron gain by the reduction process.

19. (a)

−1 loses 1 e− 0
0 gains 1 e− −1

$Br_{2(l)} + 2\ NaI_{(aq)} \rightarrow 2\ NaBr_{(aq)} + I_{2(s)}$

(b)

0 gains 2 e− −2
−2 loses 6 e− +4

$2\ PbS_{(s)} + 3\ O_{2(g)} \rightarrow 2\ PbO_{(s)} + 2\ SO_{2(g)}$

21. (a)

−1 loses 1 e− 0
+7 gains 5 e− +2

$2\ MnO_4^-{}_{(aq)} + 10\ I^-{}_{(aq)} + 16\ H^+{}_{(aq)} \rightarrow 2\ Mn^{2+}{}_{(aq)} + 5\ I_{2(s)} + 8\ H_2O_{(l)}$

(b)

+6 gains 2 e− +4
0 loses 2 e− +2

$Cu_{(s)} + 4\ H^+{}_{(aq)} + SO_4^{2-}{}_{(aq)} \rightarrow Cu^{2+}{}_{(aq)} + SO_{2(g)} + 2\ H_2O_{(l)}$

Section 17.4 *Balancing Redox Equations: Half-Reaction Method*

23. (a) $H_2O_2(aq) + 2\,H^+(aq) + 2\,e^- \rightarrow 2\,H_2O(l)$
    (b) $AsO_3^{3-}(aq) \rightarrow AsO_3^-(aq) + 2\,e^-$

25. (a) Oxidation: $3\,(Zn \rightarrow Zn^{2+} + 2\,e^-)$
    Reduction: $2\,(NO_3^- + 4\,H^+ + 3\,e^- \rightarrow NO + 2\,H_2O)$

    $3\,Zn(s) + 2\,NO_3^-(aq) + 8\,H^+(aq) \rightarrow 3\,Zn^{2+}(aq) + 2\,NO(g) + 4\,H_2O(l)$

    (b) Oxidation: $2\,(4\,H_2O + Mn^{2+} \rightarrow MnO_4^- + 8\,H^+ + 5\,e^-)$
    Reduction: $5\,(BiO_3^- + 6\,H^+ + 2\,e^- \rightarrow Bi^{3+} + 3\,H_2O)$

    $2\,Mn^{2+}(aq) + 5\,BiO_3^-(aq) + 14\,H^+(aq) \rightarrow 2\,MnO_4^-(aq) + 5\,Bi^{3+}(aq) + 7\,H_2O(l)$

27. Oxidation: $Cl_2 + 2\,H_2O \rightarrow 2\,HOCl + 2\,H^+ + 2\,e^-$
    Reduction: $Cl_2 + 2\,e^- \rightarrow 2\,Cl^-$

    $Cl_2(g) + H_2O(l) \rightarrow Cl^-(aq) + HOCl(aq) + H^+(aq)$

Section 17.5 *Predicting Spontaneous Redox Reactions*

29.

	Substances	Greater Tendency to be Oxidized
(a)	Li(*s*) or K(*s*)	Li(*s*)
(b)	Al(*s*) or Mg(*s*)	Mg(*s*)
(c)	Fe^{2+}(*aq*) or I^-(*aq*)	I^-(*aq*)
(d)	Br^-(*aq*) or Cl^-(*aq*)	Br^-(*aq*)

31.

	Substances	Stronger Reducing Agent
(a)	Cu(*s*) or Cr(*s*)	Cr(*s*)
(b)	H_2(*g*) or Cu(*s*)	H_2(*g*)
(c)	Cu(*s*) or I^-(*aq*)	Cu(*s*)
(d)	Cl^-(*aq*) or H_2(*g*)	H_2(*g*)

33. (a) nonspontaneous (b) spontaneous

35. (a) spontaneous (b) spontaneous

Section 17.6 *Voltaic Cells*

37.

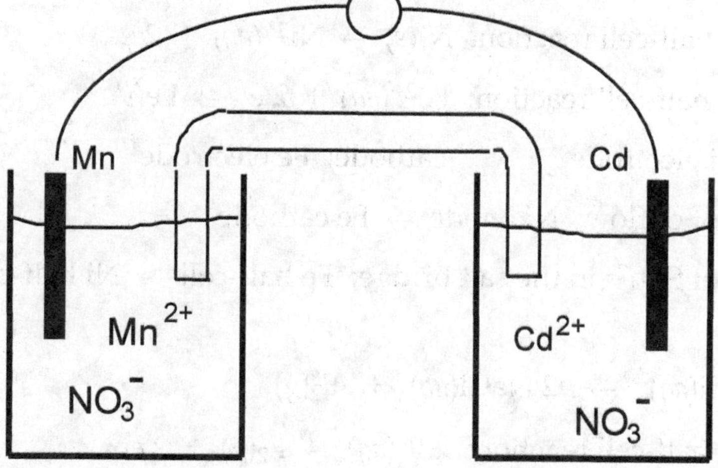

Anode: Mn Cathode: Cd

$Mn(s) + Cd(NO_3)_2(aq) \rightarrow Cd(s) + Mn(NO_3)_2(aq)$

39.

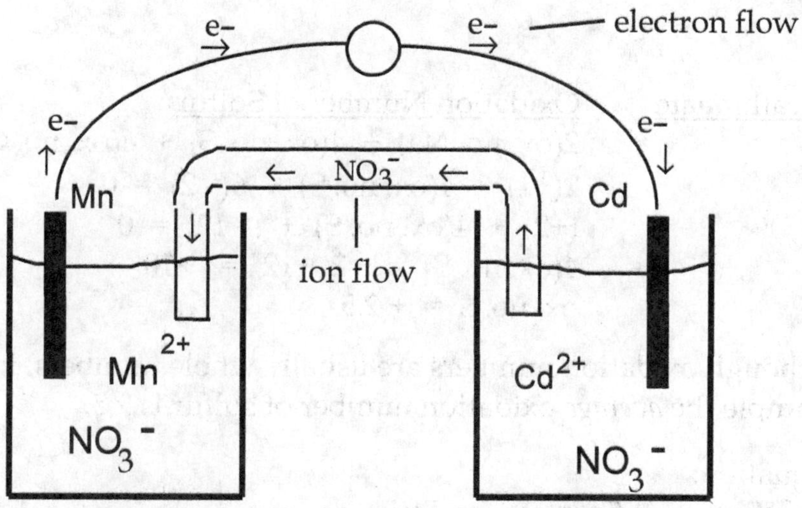

$Mn(s) + Cd(NO_3)_2(aq) \rightarrow Cd(s) + Mn(NO_3)_2(aq)$

Section 17.7 *Electrolytic Cells*

41. $Ni(s)$ + $FeSO_4(aq)$ → $Fe(s)$ + $NiSO_4(aq)$

(a) oxidation half-cell reaction: $Ni(s) \rightarrow Ni^{2+}(aq) + 2\,e^-$

(b) reduction half-cell reaction: $Fe^{2+}(aq) + 2\,e^- \rightarrow Fe(s)$

(c) anode: Ni electrode cathode: Fe electrode

(d) direction of e^- flow: Ni anode → Fe cathode

(e) direction of SO_4^{2-} in the salt bridge: Fe half-cell → Ni half-cell

43. $Cl_2(l)$ + $2\,NaF(aq)$ → $2\,NaCl(aq)$ + $F_2(g)$

(a) oxidation half-cell reaction: $2\,F^-(aq) \rightarrow F_2(g) + 2\,e^-$

(b) reduction half-cell reaction: $Cl_2(g) + 2\,e^- \rightarrow 2\,Cl^-(aq)$

(c) anode: $Pt(F_2)$ electrode cathode: $Pt(Cl_2)$ electrode

(d) direction of e^- flow: $Pt(F_2)$ anode → $Pt(Cl_2)$ cathode

(e) direction of Na^+ in the salt bridge: $Pt(F_2)$ half-cell → $Pt(Cl_2)$ half-cell

General Exercises

45. <u>Sodium Tetrathionate</u>
$Na_2S_4O_6$

<u>Oxidation Number of Sulfur</u>

$2(\text{ox. no. Na}) + 4(\text{ox. no. S}) + 6(\text{ox. no. O}) = 0$

$2(+1) + 4(\text{ox. no. S}) + 6(-2) = 0$

$(+2) + 4(\text{ox. no. S}) + (-12) = 0$

$4(\text{ox. no. S}) = -2 + 12 = +10$

$\text{ox. no. S} = +2.5$

Note: Although oxidation numbers are usually whole numbers, in this example the *average* oxidation number of sulfur is 2.5.

47. Net ionic equation:
$Co(s) + Hg^{2+}(aq) \rightarrow Co^{2+}(aq) + Hg(l)$

49. Net ionic equation:
$Zn(s) + 2\,H^+(aq) \rightarrow Zn^{2+}(aq) + H_2(g)$

Challenge Exercises

51. Net ionic equation:
$3\,Cu(s) + 8\,H^+(aq) + 2\,NO_3^-(aq) \rightarrow 3\,Cu^{2+}(aq) + 2\,NO(g) + 4\,H_2O(l)$

CHAPTER

Nuclear Chemistry 18

Section 18.1 *Natural Radioactivity*

1. An alpha particle (α) is identical to a helium nucleus, which is deflected toward the negative electrode as it passes between electrically charged plates.

3. A beta particle (β) is identical to an electron, which is deflected toward the positive electrode as it passes between electrically charged plates.

5. A gamma ray (γ) is powerful radiant energy, which is *not* deflected as it passes between electrically charged plates

Section 18.2 *Nuclear Equations*

7.

	Radiation	Approximate Mass
(a)	alpha particle (α)	~4 amu
(b)	beta particle (β^-)	~0 amu
(c)	gamma ray (γ)	0 amu
(d)	positron (β^+)	~0 amu
(e)	neutron (n^0)	~1 amu
(f)	proton (p^+)	~1 amu

9. (a) $^{238}_{92}\text{U} \rightarrow\ ^{234}_{90}\text{Th} +\ ^{4}_{2}\text{He}$

(b) $^{28}_{13}\text{Al} \rightarrow\ ^{28}_{14}\text{Si} +\ ^{0}_{-1}\text{e}$

(c) $^{15}_{8}\text{O} \rightarrow\ ^{15}_{7}\text{N} +\ ^{0}_{+1}\text{e}$

(d) $^{55}_{26}\text{Fe} +\ ^{0}_{-1}\text{e} \rightarrow\ ^{55}_{25}\text{Mn}$

11. (a) $^{221}_{88}X \rightarrow\ ^{217}_{86}Rn +\ ^{4}_{2}He$ $\qquad ^{221}_{88}X =\ ^{221}_{88}Ra$

(b) $^{43}_{19}X \rightarrow\ ^{43}_{20}Ca +\ ^{0}_{-1}e$ $\qquad ^{43}_{19}X =\ ^{43}_{19}K$

(c) $^{73}_{36}X \rightarrow\ ^{73}_{35}Br +\ ^{0}_{+1}e$ $\qquad ^{73}_{36}X =\ ^{73}_{36}Kr$

(d) $^{133}_{56}X +\ ^{0}_{-1}e \rightarrow\ ^{133}_{55}Cs$ $\qquad ^{133}_{56}X =\ ^{133}_{56}Ba$

Section 18.3 *Radioactive Decay Series*

13. $^{238}_{92}U \rightarrow\ ^{234}_{90}Th +\ ^{4}_{2}He$

$^{234}_{90}Th \rightarrow\ ^{234}_{91}Pa +\ ^{0}_{-1}e$ $\qquad$ The resulting nuclide is: $^{234}_{91}Pa$.

15. $^{235}_{92}U \rightarrow\ ^{231}_{90}Th +\ ^{4}_{2}He$

$^{231}_{90}Th \rightarrow\ ^{231}_{91}Pa +\ ^{0}_{-1}e$ $\qquad$ The resulting nuclide is: $^{231}_{91}Pa$.

17. <u>Decay Series for Thorium-232</u>

$^{232}_{90}Th \xrightarrow{\alpha}\ ^{228}_{88}Ra \xrightarrow{\beta}\ ^{228}_{89}Ac$

$^{228}_{89}Ac \xrightarrow{\beta}\ ^{228}_{90}Th$

$^{228}_{90}Th \xrightarrow{\alpha}\ ^{224}_{88}Ra \xrightarrow{\alpha}\ ^{220}_{86}Rn$

$^{220}_{86}Rn \xrightarrow{\alpha}\ ^{216}_{84}Po$

$^{216}_{84}Po \xrightarrow{\alpha}\ ^{212}_{82}Pb \xrightarrow{\beta}\ ^{212}_{83}Bi$

$^{212}_{83}Bi \xrightarrow{\beta}\ ^{212}_{84}Po$

$^{212}_{84}Po \xrightarrow{\alpha}\ ^{208}_{82}Pb$

<u>Answers</u>

(a)	α	(b)	β
(c)	β	(d)	α
(e)	α	(f)	α
(g)	α	(h)	β
(i)	β	(j)	α

Section 18.4 *Radioactive Half-Life*

19. $100\% \times \frac{1}{2} \times \frac{1}{2} = 25\%$

25% of the U-238 radionuclides remain after two half-lives.

21.

Activity of C-14	Elapsed Time
240 dpm	$0\ t_{1/2}$
120 dpm	$1\ t_{1/2}$
60 dpm	$2\ t_{1/2}$

$$2\ \cancel{t_{1/2}} \times \frac{5730 \text{ years}}{1\ \cancel{t_{1/2}}} \approx 11{,}500 \text{ years}$$

23. $60\ \cancel{\text{hours}} \times \frac{1\ t_{1/2}}{15\ \cancel{\text{hours}}} = 4\ t_{1/2}$

$$80 \text{ mg Na-24} \times \frac{1}{2} \times \frac{1}{2} \times \frac{1}{2} \times \frac{1}{2} = 5 \text{ mg Na-24}$$

Section 18.5 *Applications of Radionuclides*

25. Radiocarbon dating (i.e., carbon-14) can be used to date fossils, which are up to 50,000 years old.

27. The γ-emitting radionuclide cobalt–60 can be used to irradiate food and destroy microorganisms.

29. The γ-emitting radionuclide technetium-99 can be used to diagnose and locate brain tumors.

Section 18.6 *Induced Radioactivity*

31. ${}^{23}_{11}\text{Na} + {}^{1}_{1}\text{H} \rightarrow {}^{23}_{12}\text{Mg} + {}^{1}_{0}\text{n}$ The radionuclide is: ${}^{23}_{12}\text{Mg}$.

33. ${}^{81}_{35}\text{Br} + {}^{0}_{0}\gamma \rightarrow {}^{80}_{35}\text{Br} + {}^{1}_{0}\text{n}$ The radionuclide is: ${}^{80}_{35}\text{Br}$.

35. ${}^{243}_{95}\text{Am} + {}^{22}_{10}\text{Ne} \rightarrow {}^{260}_{105}\text{Db} + 5\,{}^{1}_{0}\text{n}$ The Russians created: ${}^{260}_{105}\text{Db}$.

37. ${}^{54}_{24}\text{Cr} + {}^{207}_{82}\text{Pb} \rightarrow {}^{260}_{106}\text{Sg} + {}^{1}_{0}\text{n}$ The Russians created: ${}^{260}_{106}\text{Sg}$.

Section 18.7 *Nuclear Fission*

39. $^{1}_{0}n \xrightarrow{\text{1st}} 2\,^{1}_{0}n \xrightarrow{\text{2nd}} 4\,^{1}_{0}n \xrightarrow{\text{3rd}} 8\,^{1}_{0}n$

After the third step, 8 neutrons are produced.

41. Uranium must be enriched before it can be used in the fuel rods of a nuclear reactor because naturally occurring uranium contains only 0.7% *fissionable* U-235 and 99.3% *nonfissionable* U-238.

43. $^{235}_{92}U + ^{1}_{0}n \rightarrow ^{144}_{54}Xe + ^{90}_{38}Sr + 2\,^{1}_{0}n$ The number of neutrons is 2.

45. $^{233}_{92}U + ^{1}_{0}n \rightarrow ^{143}_{56}Ba + ^{88}_{36}Kr + 3\,^{1}_{0}n$ The unknown nuclide is: $^{233}_{92}U$.

Section 18.8 *Nuclear Fusion*

47. $^{2}_{1}H + ^{2}_{1}H \rightarrow ^{3}_{1}H + ^{0}_{+1}e + ^{1}_{0}X$ The particle *X* is: $^{1}_{0}n$.

49. $^{3}_{2}He + ^{1}_{1}X \rightarrow ^{4}_{2}He + ^{0}_{+1}e$ The particle *X* is: $^{1}_{1}H$.

51. $^{2}_{1}X + ^{2}_{1}X \rightarrow ^{4}_{2}He + ^{0}_{0}\gamma$ The unknown nuclide is: $^{2}_{1}H$.

General Exercises

53. Since an activity of ~7.7 dpm per gram indicates that one half-life has expired, the age of Crater Lake is ~6000 years because 1 $t_{1/2}$ = 5730 years.

55. $^{241}_{95}X \rightarrow ^{237}_{93}Np + ^{4}_{2}He$ The radionuclide is: $^{241}_{95}Am$.

Challenge Exercises

57. $^{2}_{1}H + ^{3}_{1}H \rightarrow ^{4}_{2}He + ^{1}_{0}X$ + energy The emitted particle is: $^{1}_{0}n$.

CHAPTER

19

Organic Chemistry

Section 19.1 *Hydrocarbons*

1. The alkane class of compounds consists of saturated hydrocarbons.

3. The alkane class of compounds has only single bonds.

5. The alkyne class of compounds has a triple bond.

Section 19.2 *Alkanes*

7.

	Molecular Formula	Class of Compound
(a)	C_6H_{14}	The molecular formula of the alkanes is C_nH_{2n+2}. Given C_6H_{14} fits the general formula, it is an alkane.
(b)	C_6H_{12}	The molecular formula of the alkanes is C_nH_{2n+2}. Given C_6H_{12} does *not* fit the general formula, it is *not* an alkane.

9.

	Structural Formula	IUPAC Name
(a)	CH_3-CH_2-CH_2-CH_3	butane
(b)	CH_3-CH_2-CH_2-CH_2-CH_2-CH_3	hexane
(c)	CH_3-CH_2-CH_2-CH_2-CH_2-CH_2-CH_2-CH_3	octane
(d)	CH_3-CH_2-CH_2-CH_2-CH_2-CH_2-CH_2-CH_2-CH_2-CH_3	decane

11. Three Isomers of Pentane (C_5H_{12})

```
                                                              CH3
                                                               |
CH3-CH2-CH2-CH2-CH3     CH3-CH2-CH-CH3           CH3 - C - CH3
                                 |                             |
                                CH3                           CH3
```

13. Two Isomers of Bromopropane (C_3H_7Br)

CH_3-CH_2-CH_2-Br $\qquad$ CH_3-CH(Br)-CH_3

15.

	Alkyl Group	IUPAC Name
(a)	methyl	CH_3-
(b)	ethyl	CH_3CH_2-

17. (a) 2-methylpentane (b) 4-ethyl-2-methylhexane
(c) 2,4,4-trimethylheptane (d) 3,4,5-trimethyloctane

19. (a) $CH_4 + 2\,O_2 \rightarrow CO_2 + 2\,H_2O$
(b) $C_3H_8 + 5\,O_2 \rightarrow 3\,CO_2 + 4\,H_2O$
(c) $2\,C_6H_{14} + 19\,O_2 \rightarrow 12\,CO_2 + 14\,H_2O$
(d) $2\,C_8H_{18} + 25\,O_2 \rightarrow 16\,CO_2 + 18\,H_2O$

Section 19.3 *Alkenes and Alkynes*

21.

	Molecular Formula	Class of Compound
(a)	C_6H_{14}	The molecular formula of the alkenes is C_nH_{2n}. Given C_6H_{14} does *not* fit the general formula, it is *not* an alkene.
(b)	C_6H_{12}	The molecular formula of the alkenes is C_nH_{2n}. Given C_6H_{12} fits the general formula, it is an alkene.

23.

	Structural Formula	IUPAC Name
(a)	CH_2=CH-CH_2-CH_3	1-butene
(b)	CH_2=CH-CH_2-CH_2-CH_3	1-pentene
(c)	CH_3-CH=CH-CH_2-CH_2-CH_3	2-hexene
(d)	CH_3-CH_2-CH_2-CH=CH-CH_2-CH_2-CH_3	4-octene

25.

	Molecular Formula	Class of Compound
(a)	C_6H_{12}	The molecular formula of the alkynes is C_nH_{2n-2}. Given C_6H_{12} does *not* fit the general formula, it is *not* an alkyne.
(b)	C_6H_{10}	The molecular formula of the alkynes is C_nH_{2n-2}. Given C_6H_{10} fits the general formula, it is an alkyne.

27.

	Structural Formula	IUPAC Name
(a)	$CH_3\text{-}C{\equiv}C\text{-}CH_2\text{-}CH_3$	2-pentyne
(b)	$CH{\equiv}C\text{-}CH_2\text{-}CH_2\text{-}CH_3$	1-pentyne
(c)	$CH_3\text{-}CH_2\text{-}C{\equiv}C\text{-}CH_2\text{-}CH_3$	3-hexyne
(d)	$CH_3\text{-}C{\equiv}C\text{-}CH_2\text{-}CH_2\text{-}CH_2\text{-}CH_3$	2-heptyne

29. Two Isomers of Straight-Chain Pentene (C_5H_{10})

$CH_2{=}CH\text{-}CH_2\text{-}CH_2\text{-}CH_3$	$CH_3\text{-}CH{=}CH\text{-}CH_2\text{-}CH_3$
1-pentene	2-pentene

31. (a)

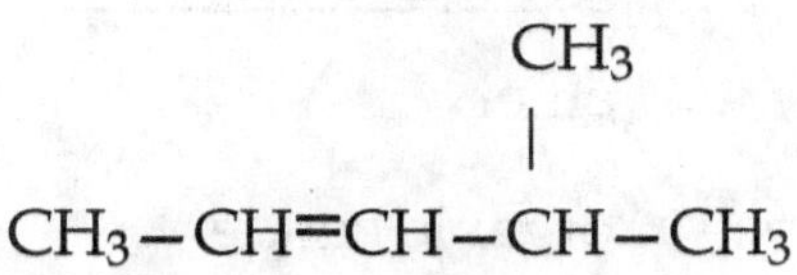

4-methyl-2-pentene

(b)

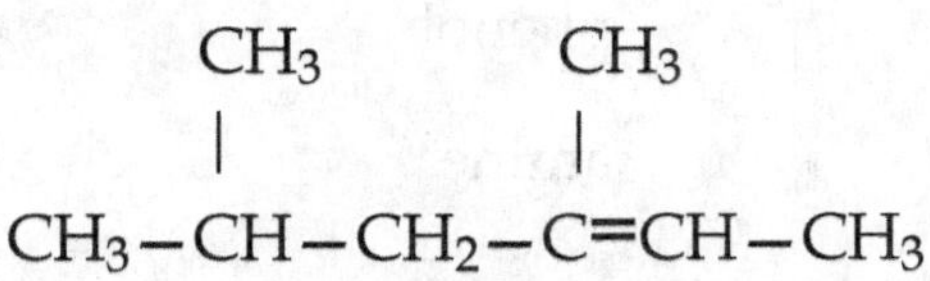

3,5-dimethyl-2-hexene

33. (a)

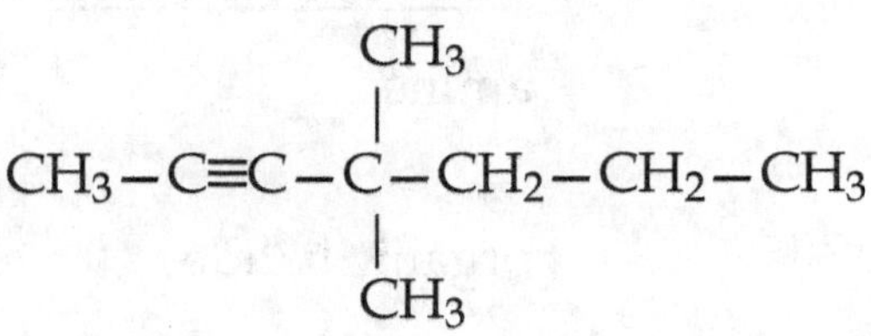

4,4-dimethyl-2-heptyne

(b)

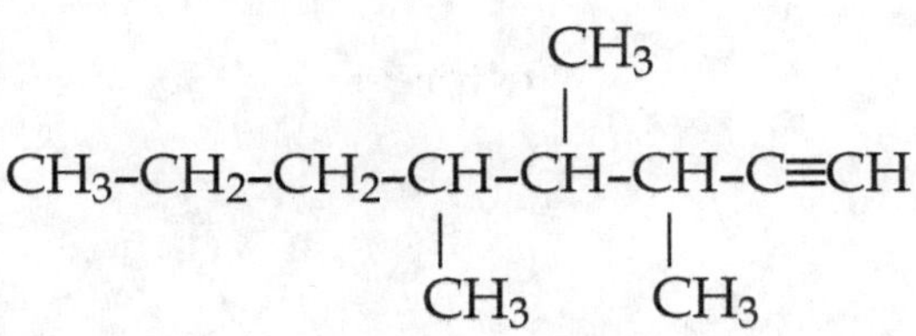

3,4,5-trimethyl-1-octyne

35. (a) $CH_2{=}CH_2 + 3\,O_2 \rightarrow 2\,CO_2 + 2\,H_2O$

(b) $CH_3\text{-}CH{=}CH_2 + H_2 \rightarrow CH_3\text{-}CH_2\text{-}CH_3$

Section 19.4 *Arenes*

37. <u>Three Isomers of Xylene</u> $[C_6H_4(CH_3)_2]$

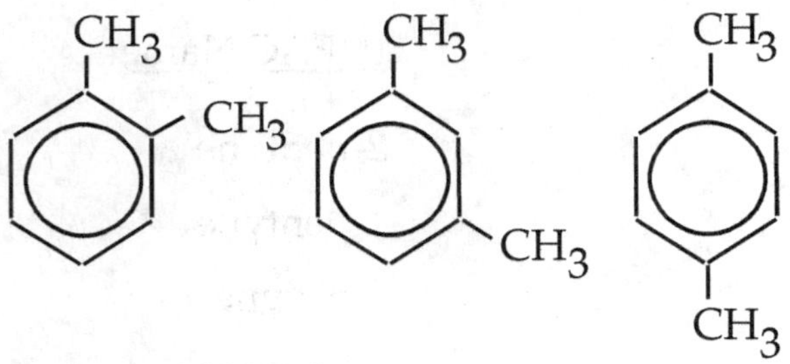

ortho- *meta-* *para-*

39. $C_6H_6 + O_2 \xrightarrow{\text{spark}} CO_2 + H_2O$ (carbon dioxide and water)

Section 19.5 *Hydrocarbon Derivatives*

41.

	General Formula	Class of Compound
(a)	R–O–R′	ether
(b)	R–X	organic halide
(c)	Ar–OH	phenol
(d)	$R{-}NH_2$	amine

43.

	Chemical Formula	Class of Compound
(a)	$CH_3\text{-}NH_2$	amine
(b)	$CH_3CH_2\text{-}F$	organic halide
(c)	$CH_3{-}\overset{\overset{\large O}{\|\|}}{C}{-}O{-}CH_3$	ester
(d)	$H{-}\overset{\overset{\large O}{\|\|}}{C}{-}OH$	carboxylic acid

(e) 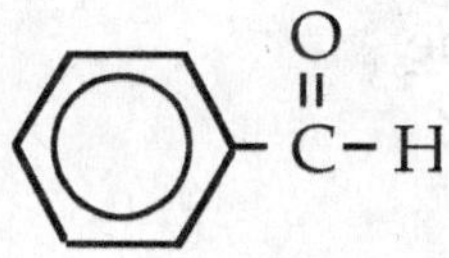 aldehyde

(f) 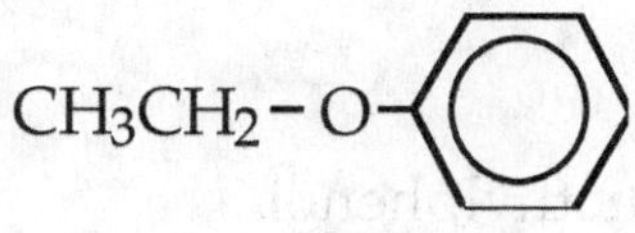 ether

(g) 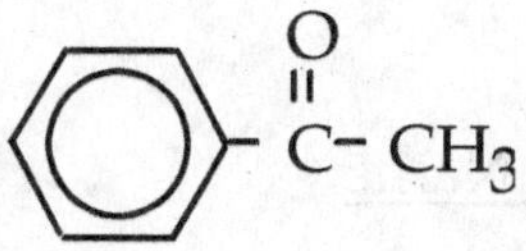 ketone

(h) 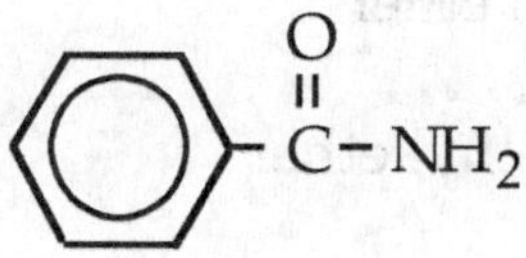 amide

Section 19.6 *Organic Halides*

45.

	Organic Halide	IUPAC Name	Common Name
(a)	CH_3-I	iodomethane	"methyl iodide"
(b)	CH_3-CH_2-Br	bromoethane	"ethyl bromide"
(c)	CH_3-CH_2-CH_2-F	1-fluoropropane	"propyl fluoride"
(d)	$(CH_3)_2$-CH-Cl	2-chloropropane	"isopropyl chloride"

47. 1,1,2-trichloroethylene CCl_2=CHCl

Section 19.7 *Alcohols, Phenols, and Ethers*

49.

	Alcohol	IUPAC Name
(a)	CH_3-CH_2-CH_2-CH_2-OH	1-butanol
(b)	CH_3-CH_2-CH(OH)-CH_3	2-butanol
(c)	CH_2(OH)-CH_2-CH_2-CH_3	1-butanol
(d)	CH_3-CH(OH)-CH_2-CH_3	2-butanol

51.

	Phenol	IUPAC Name
(a)	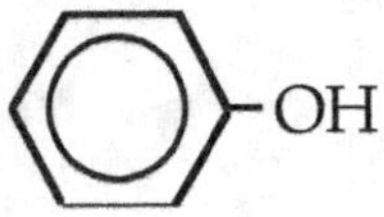	phenol
(b)		*para*-methylphenol

53.

	Ether	Common Name
(a)	CH_3-O-CH_3	"dimethyl ether"
(b)	CH_3-O-CH_2CH_3	"methyl ethyl ether"
(c)	$CH_3CH_2CH_2$-O-$CH_2CH_2CH_3$	"dipropyl ether"
(d)	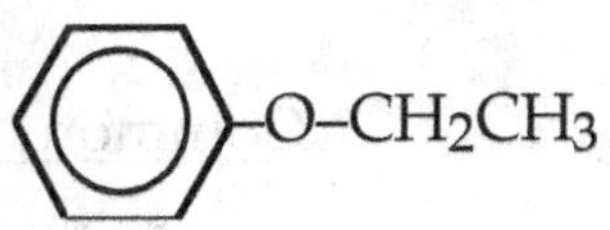	"phenyl ethyl ether"

55. Although "ethyl alcohol" and "dimethyl ether" have the same molecular mass, the alcohol has a higher boiling point. Alcohols have strong intermolecular attraction due to hydrogen bonding, whereas ethers have weak dipole attraction.

57.

Alcohol (C_2H_6O)	Ether (C_2H_6O)
CH_3-CH_2-OH	CH_3-O-CH_3

Section 19.8 *Amines*

59.

	Amine	IUPAC Name	Common Name
(a)	CH_3-CH_2-NH_2	amino ethane	"ethylamine"
(b)	CH_3-CH_2-CH_2-NH_2	1-amino propane	"propylamine"

61. Although "ethylamine" and propane have about the same molecular mass, "ethylamine" has a higher boiling point. Amines have a strong intermolecular attraction due to hydrogen bonding, while alkanes have a weak intermolecular attraction due to dispersion forces.

Section 19.9 *Aldehydes and Ketones*

63.

	Aldehyde	IUPAC Name	Common Name
(a)	$H-\overset{O}{\overset{\|}{C}}-H$	methanal	"formaldehyde"
(b)	$CH_3-\overset{O}{\overset{\|}{C}}-H$	ethanal	"acetaldehyde"

65.

	Ketone	IUPAC Name	Common Name
(a)	$CH_3-\overset{O}{\overset{\|}{C}}-CH_3$	2-propanone	"dimethyl ketone"
(b)	$CH_3-\overset{O}{\overset{\|}{C}}-CH_2CH_3$	2-butanone	"methyl ethyl ketone"

67.

Aldehyde (C_3H_6O)	Ketone (C_3H_6O)
$CH_3CH_2-\overset{O}{\overset{\|}{C}}-H$	$CH_3-\overset{O}{\overset{\|}{C}}-CH_3$

Section 19.10 *Carboxylic Acids, Esters, and Amides*

69.

	Carboxylic Acid	IUPAC Name	Common Name
(a)	$H-\overset{O}{\overset{\|}{C}}-OH$	methanoic acid	"formic acid"
(b)	$CH_3-\overset{O}{\overset{\|}{C}}-OH$	ethanoic acid	"acetic acid"

71.

	Ester	IUPAC Name	Common Name
(a)	$H-\overset{O}{\overset{\|}{C}}-O-CH_2CH_3$	ethyl methanoate	"ethyl formate"
(b)	$CH_3CH_2-\overset{O}{\overset{\|}{C}}-O-CH_2CH_3$	ethyl propanoate	"ethyl propionate"

73. (a) $CH_3\text{-}C(=O)\text{-}OH + CH_3\text{-}OH \xrightarrow{H_2SO_4} CH_3\text{-}C(=O)\text{-}O\text{-}CH_3 + H_2O$

(b)

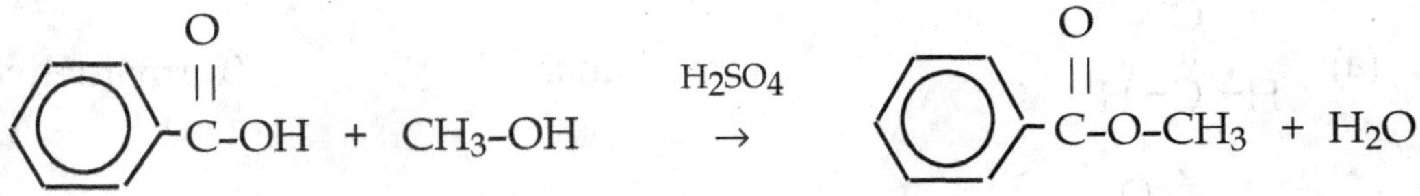

75.

	Amide	IUPAC Name	Common Name
(a)	$H\text{-}C(=O)\text{-}NH_2$	methanamide	"formamide"
(b)	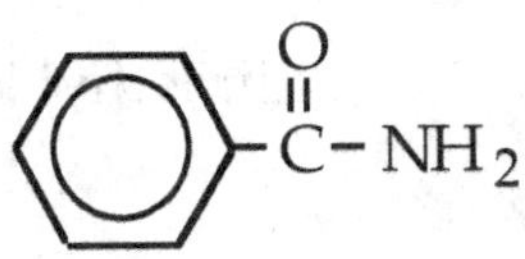	benzamide	–

General Exercises

77.

	Compound	Classification
(a)	eicosane	saturated (*–ane* suffix)
(b)	dodecene	unsaturated (*–ene* suffix)

79. Acetylsalicylic acid has an *–ic* acid suffix; thus, the functional group present is a *carboxylic acid.*

81. Androsterone has an *–one* suffix; thus, the functional group present is a *ketone.*

83. Chloral hydrate has a *chlor–* prefix and has an *–al* suffix; thus, the functional groups present are an *organic chloride* and an *aldehyde.*

Challenge Exercises

85.

	Compounds	Higher Boiling Point
(a)	$CH_3\text{-}O\text{-}CH_3$ or $CH_3CH_2\text{-}OH$	$CH_3CH_2\text{-}OH$
(b)	$CH_3CH_2CH_2\text{-}NH_2$ or $CH_3CH_2CH_2\text{-}F$	$CH_3CH_2CH_2\text{-}NH_2$

In each of the above pairs, the compound with the higher boiling point can hydrogen bond. (a) $CH_3CH_2\text{-}OH$ has an -OH and can hydrogen bond; $CH_3\text{-}O\text{-}CH_3$ cannot. (b) $CH_3CH_2CH_2\text{-}NH_2$ has an $-NH_2$ and can hydrogen bond; $CH_3CH_2CH_2\text{-}F$ cannot.

Hydrogen bonds increase the effective molar mass of the compound in the liquid, and raise the boiling point by increasing the energy required to separate molecules as they leave the liquid and enter the vapor state.

CHAPTER

Biochemistry 20

Section 20.1 *Biological Compounds*

1. (a) A *protein* is a polymer consisting of many amino acids.
   (b) A *carbohydrate* is a compound consisting of simple sugar molecules.

3. (a) Two amino acids in a protein are joined by a *peptide linkage*.
   (b) Two sugar molecules in a carbohydrate are joined by a *glycoside linkage*.

Section 20.2 *Proteins*

5. A covalent amide bond is responsible for linking two amino acids together in the *primary structure* of a protein.

7. The *primary structure* of a protein refers to the amino acid sequence. The *secondary structure* of a protein refers to the shape of the chain, for example, an α-helix.

9. The dipeptide formed from two molecules of alanine is named alanylalanine (Ala–Ala):

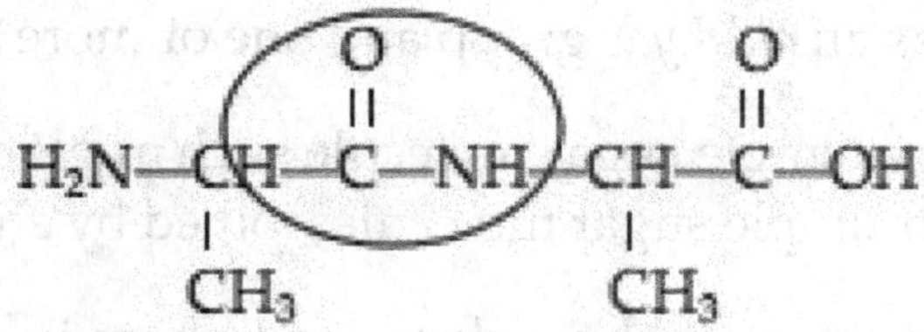

11. The dipeptide formed from a molecule of serine and a molecule of tyrosine is named seryltyrosine (Ser–Tyr):

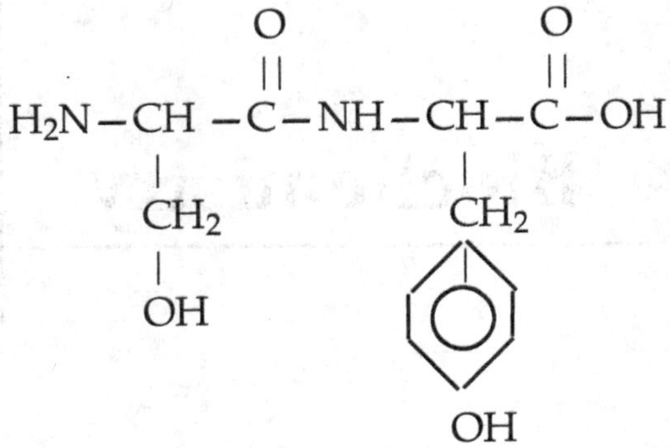

13.

Tripeptide	Amino Acid Sequence
1	Ala–Gln–Met
2	Ala–Met–Gln
3	Gln–Ala–Met
4	Gln–Met–Ala
5	Met–Ala–Gln
6	Met–Gln–Ala

15. The three amino acids in the segment of protein chain are *glutamic acid, lysine,* and *serine*.

17. Two of the amino acids in oxytocin differ from those in vasopressin. In oxytocin, *isoleucine* replaces phenylalanine, and *leucine* replaces arginine.

Section 20.3 *Enzymes*

19. In the lock-and-key model, the key represents an *enzyme (E)*.

21. In the lock-and-key model, the "teeth" on the key represent an active site on the enzyme molecule.

23. An enzyme is likely to be inactive under experimental conditions where the pH or temperature is too high or too low.

Section 20.4 *Carbohydrates*

25. A pentose sugar molecule has five carbon atoms.

27. An aldose sugar has an *aldehyde* group and one or more *alcohol* groups.

29. A *monosaccharide* is a simple sugar molecule such as glucose, whereas a *disaccharide* has two simple sugar molecules joined by a glycoside linkage.

31. *Glucose* and *fructose* are joined by a glycoside linkage in sucrose.

Section 20.5 *Lipids*

33.

	Characteristic	Lipid
(a)	animal source	fat
(b)	plant source	oil
(c)	semisolid	fat
(d)	insoluble in water	fat and oil

35.

	Name of Fatty Acid	Structure of Fatty Acid
(a)	palmitic acid	$HOOC\text{-}(CH_2)_{14}\text{-}CH_3$
(b)	lauric acid	$HOOC\text{-}(CH_2)_{10}\text{-}CH_3$
(c)	myristic acid	$HOOC\text{-}(CH_2)_{12}\text{-}CH_3$

37. A phospholipid has *two* fatty acids and phosphoric acid esterifed to glycerol; whereas, a triglyceride (fat, oil) has *three* fatty acids esterified to glycerol.

39. The saponification of carnauba wax with aqueous NaOH produces the sodium salt of a fatty acid and long-chain alcohol.

$$CH_3\text{-}(CH_2)_{24}\text{-}\overset{\overset{\large O}{||}}{C}\text{-}O\text{-}(CH_2)_{29}\text{-}CH_3 \;+\; NaOH \;\rightarrow$$

$$CH_3\text{-}(CH_2)_{24}\text{-}\overset{\overset{\large O}{||}}{C}\text{-}O^-\ Na^+ \;+\; CH_3\text{-}(CH_2)_{29}\text{-}OH$$

41.

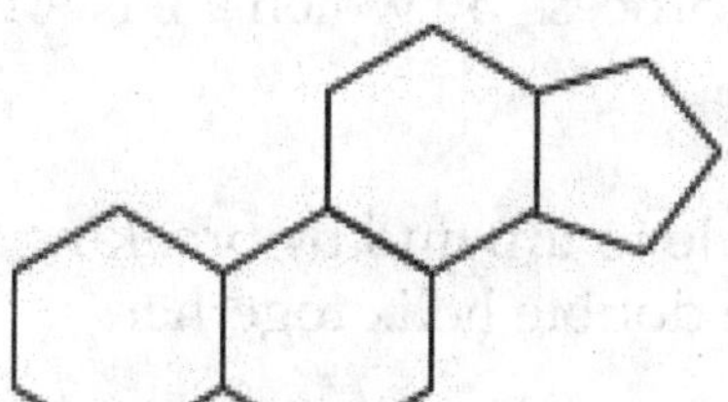

Section 20.6 *Nucleic Acids*

43. The three general components of a DNA nucleotide are deoxyribose sugar, a nitrogen base, and phosphoric acid.

45. The four DNA bases are adenine (A), cytosine (C), guanine (G), and thymine (T).

47. A DNA molecule has two strands of nucleotides joined by hydrogen bonds.

49. There are two hydrogen bonds between adenine and thymine (A=T).

51. During DNA replication, an adenine base (A) on the template strand codes for a thymine base (T) base on the complementary strand.

53. During RNA transcription, an adenine base (A) on the template DNA strand codes for a uracil base (U) on the growing RNA strand.

General Exercises

55. A protein that provides strength, such as that in muscle tissue, has a long and extended shape.

57. When a protein is denatured in aqueous acid hydrogen bonds are broken, which form the secondary and tertiary structure, and the molecule assumes a random shape.

59. The water-insoluble vitamins (A, D, E, K) are lipids.

61. There are six possible codons having an adenine base (A), a cytosine base (C), and a guanine base (G): ACG, AGC, CAG, CGA, GAC, and GCA.

Challenge Exercises

63. *Replication* is the process by which a DNA molecule synthesizes an identical molecule of DNA; whereas, *transcription* is the process by which a DNA molecule synthesizes a complementary strand of RNA.

65. The first step in *replication* is for a DNA molecule to unwind by breaking the hydrogen bonds (A=T and C≡G) that hold the double helix together.

In the second step, each strand of DNA acts as a template to synthesize a complementary strand of DNA. That is, a nucleotide in the template strand that contains the base adenine (A) codes for thymine (T) on the growing strand; cytosine (C) codes for guanine (G); guanine (G) codes for cytosine (C); and thymine (T) codes for adenine (A).

At the conclusion, each pair of template strands and complementary strands are joined by hydrogen bonds (A=T and C≡G) and the two resulting structures are exact duplicates of the original DNA molecule. Thus, two identical molecules of DNA are produced from one molecule of DNA.

Flashcards

for

Common Monoatomic and Polyatomic

Cations and Anions

barium ion

magnesium ion

strontium ion

calcium ion

potassium ion

sodium ion

lithium ion

hydrogen ion

Mg^{2+}

Ba^{2+}

Ca^{2+}

Sr^{2+}

Na^{+}

K^{+}

H^{+}

Li^{+}

manganese(II) ion

aluminum ion

chromium(III) ion

nickel(II) ion

cadmium ion

silver ion

zinc ion

Al^{3+}

Mn^{2+}

Ni^{2+}

Cr^{3+}

Ag^{+}

Cd^{2+}

Zn^{2+}

copper(II) ion
cupric ion

copper(I) ion
cuprous ion

iron(III) ion
ferric ion

iron(II) ion
ferrous ion

lead(IV) ion
plumbic ion

lead(II) ion
plumbous ion

tin(IV) ion
stannic ion

tin(II) ion
stannous ion

Cu^{+}	Cu^{2+}
Fe^{2+}	Fe^{3+}
Pb^{2+}	Pb^{4+}
Sn^{2+}	Sn^{4+}

cobalt(III) ion
cobaltic ion

cobalt(II) ion
cobaltous ion

mercury(II) ion
mercuric ion

mercury(I) ion
mercurous ion

hydronium ion

ammonium ion

hydroxide ion

cyanide ion

Co^{2+} | Co^{3+}

Hg_2^{2+} | Hg^{2+}

NH_4^+ | H_3O^+

CN^- | OH^-

chloride ion

fluoride ion

iodide ion

bromide ion

sulfide ion

oxide ion

phosphide ion

nitride ion

F^{-}

Cl^{-}

Br^{-}

I^{-}

O^{2-}

S^{2-}

N^{3-}

P^{3-}

hypochlorite ion	perchlorate ion
chlorite ion	chlorate ion
nitrite ion	nitrate ion
sulfite ion	sulfate ion

ClO_4^-

ClO^-

ClO_3^-

ClO_2^-

NO_3^-

NO_2^-

SO_4^{2-}

SO_3^{2-}

hydrogen carbonate ion

carbonate ion

phosphate ion

hydrogen sulfate ion

permanganate ion

acetate ion

dichromate ion

chromate ion

CO_3^{2-}

HCO_3^{-}

HSO_4^{-}

PO_4^{3-}

$C_2H_3O_2^{-}$

MnO_4^{-}

CrO_4^{2-}

$Cr_2O_7^{2-}$